Frogs and Toads of the World

CHRIS MATTISON

Princeton University Press
Princeton and Oxford

Acknowledgments

Many people have been instrumental in the process that has led to this book and it is a pleasure to acknowledge them. Alexander Haas and Miguel Vences helped with identification of the frogs photographed in the field and Laurie Vitt provided information on the recent classification of frogs. Photographs of frogs and toads in habitat were taken on various field trips in the company of a number of people who helped to find specimens. They include, in alphabetical order, Philippe Blais, Alan Francis, Terry Howe, Nick Garbutt, Steven Krause, Jackie Manning, Anslem da Silva, Dave and Sigrun Tollerton and a number of local guides, especially in Borneo and Madagascar, where their expertise and sharp eyes were invaluable. It is a particular pleasure to mention my wife Gretchen Mattison and daughter Victoria in this context. Others located captive animals for studio photography, or allowed me to photograph specimens in their collections, and they include John Armitage, David Birkbeck, British Herpetological Supplies, David Burbage, Alan Drummond, Andrew Gray, Toby Mace, John Pickett, Wharf Aquatics (Wayne Swift and Craig Robinson), Martin Withers, Dawn and Gary Wood.

FRONT COVER: small red-eyed leaf frog (p.151)
BACK COVER FROM TOP TO BOTTOM: harlequin poison dart frog (p.30), African dwarf bullfrog (p.177), file-eared tree frog (p.9)
FLAPS: Oriental fire-bellied toad (p.61), golden mantella (p.75)
TITLE PAGE, VERSO, CONTENTS: red-headed poison dart frog (p.62), white-spotted reed frog (p.172), Spurrell's flying frog (p.24)
© Chris Mattison

Published in the United States, Canada, South America, Central America, and the Caribbean by Princeton University Press, 41 William Street, Princeton, New Jersey, 08540.

First published by the Natural History Museum, Cromwell Road, London SW7 5BD
© Natural History Museum, London, 2011 under the title *Frogs and Toads*.

nathist.press.princeton.edu

Library of Congress Control Number 2010932660
ISBN 978-0-691-14968-4
This book has been composed in Ingrid

Designed by Mercer Design, London
Reproduction by Saxon Digital Services
Printed in China by C&C Offset

Contents

Introduction

FROGS AND TOADS ARE THE MOST ABUNDANT and familiar amphibians. They were the first inhabitants of the Earth to have a true voice and they quickly made themselves known by croaking, peeping or trilling, especially in the breeding season. Many have names that describe their voices: bullfrog, pig frog, barking frog, spring peeper, blacksmith frog, and so on. Although frogs and toads produce a range of reactions among humans they tend to be seen as less threatening than lizards and snakes for instance, and frogs often figure in fairy tales and children's stories.

At the time of writing, there are 5,858 species of frogs and toads. They account for nearly 90% of all amphibians, the other orders being the newts and salamanders with 597 species and the caecilians with 183. Recent research has resulted in the recognition of many new families, which now total 49, as well as the discovery of new species. (Just to put this into perspective 3,445 species of frogs were recognized in 1983 and they were divided into 21 families, so the number of species grew by almost 65% and the number of families doubled in 17 years.) Not all of the additional species are new however, because many of the older ones have, in the light of more recent research, been found to consist of two or more species. Some families contain just a handful of species, others contain several hundred and, whereas some families are found around the world, others are restricted to very small regions.

The strange, recently discovered purple frog, *Nasikabatrachus sahyadrensis*, for example, discovered in 2003 and the sole member of a new family, occurs only in the Western Ghats of India, while the family Ceuthomantidae, only created in 2009, contains three species of small frogs that are restricted to the table-top mountains, or tepuis, of the Guiana Highlands. Several other families contain only one to ten species, relics of families that might have been more widespread in former times but which are now reduced to small populations as other species evolved and replaced them. One of the most successful families is the toad family, Bufonidae, which has more than 500 members occurring on every continent except Antarctica (although the single species in Australia has been introduced by humans). Another widespread family is the tree frogs, Hylidae, which has about 850 members scattered across many parts of the world but is most abundant in the rainforests of Central and South America.

Tropical regions are home to the greatest numbers of species. A few acres of tropical forest in Madagascar may contain 100 species, as many as the entire United States, and Madagascar as a whole may be home to 500 species, including many as yet unnamed. Other hot-spots include the tropics of Central and South America, West Africa and Sri Lanka.

With such a large number of species, it is no surprise that frogs and toads are very diverse in shape, size and colour, as well as in lifestyle. The largest species is the Goliath frog from West Africa, which can grow to 300 mm or more from its snout to the end of its back, with the largest recorded being 368 mm (14 ½ in), and can weigh as much as 3.66 kg (8 lb). The largest American species is the bullfrog, which can measure up to 910 mm (36 in) with its legs stretched out and weigh over 3 kg (6 ³/₅ lb). There are several contenders for the smallest species and many tropical frogs are fully grown at little more than 10 mm (²/₅ in). Newly-hatched Gardiner's frogs from the Seychelles are about the same size as a grain of rice.

Frogs and toads may be colourful or dull. Most of them are camouflaged, so they are hard for predators to spot. Depending on where they live, camouflaged, or cryptic, frogs are brown, grey or green. Some, such as the Asian horned frog, are even shaped and coloured like dead leaves so that they are almost invisible when they rest on the forest floor. Others are brightly coloured in shades of red, orange, yellow or even blue – these are poisonous species that use their colours as signals to warn predators that their skins contain toxic substances. The golden poison dart frog, *Phyllobates terribilis*, is the most poisonous animal known to science and the skin from a single individual contains enough poison to kill several humans.

For the most part, this vast array of frogs goes about the business of surviving and proliferating with little or no inconvenience to humans. Unfortunately, we do not show them the same consideration. Frogs across the globe are facing a crisis. Up to one-third of all species are facing extinction in the near future and many have already disappeared, while species that used to be common are becoming rare. The causes are varied and, in some cases complicated, but one thing is certain, unless something changes soon the world will be the poorer by up to 2,000 species of frogs. This would be an ecological tragedy because, as the environmentalist Paul Ehrlich said 'The first rule of intelligent tinkering is to save all the parts.'

ABOVE The harlequin flying frog, *Rhacophorus pardalis*, silhouetted on a leaf in Danum Valley, Borneo.

1 Origins and classification

FROGS AND TOADS ARE AMPHIBIANS, an important group because they were the first vertebrates to leave behind a purely aquatic lifestyle and begin the colonization of the land, about 350 million years ago. Amphibians are still strongly tied to water and have moist skins over which much of their respiration takes place, and so they tend to be most common in wetter, more humid habitats. In addition, most species need to return to water to breed. They lay eggs that lack shells and develop into aquatic larvae, or tadpoles, before eventually metamorphosing into the adult, terrestrial form. Some deviate from this pattern by laying their eggs on land; the tadpoles develop inside the egg, hatching as fully formed frogs in a process known as direct development and thereby circumventing the need to return to water to breed, but they can only do so if suitably moist sites are available. A very small number give birth to live young. Conversely, a few species are strictly aquatic and remain in the water even after they have metamorphosed.

ABOVE A typical frog tadpole, *Rana temporaria*, at an early stage in its development.

OPPOSITE A fossil of *Rana pueyoi*, an extinct frog that lived 6–9 million years ago in what is now Spain.

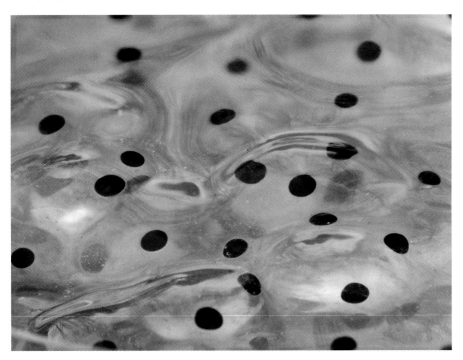

LEFT Frog spawn, such as the spawn of the frog *Rana temporaria*, consists of eggs surrounded by one or more jelly capsules and is usually laid in water.

RIGHT Caecilians are little-known amphibians restricted to tropical regions. Many are terrestrial but some, including *Typhlonectes natans* shown here, are aquatic and live in lakes and swamps.

Living amphibians are divided into three orders: Gymnophiona, the caecilians; Caudata (sometimes called the Urodela), the newts and salamanders and Anura, the frogs and toads. The caecilians are the smallest group with 183 species, and is the group about which least is known. They are elongated, limbless amphibians that live in the tropics, including seven species on the Seychelles. Most are burrowing animals, living in soil, leaf-litter and mud, but a few are aquatic. They have segmented bodies, small eyes that are easily overlooked, and moist, slimy bodies and can be mistaken for large earthworms. They are divided into three or more families and they include live-bearing and egg-laying species, but the natural history of most of the species is poorly known.

The newts and salamanders, with 597 species across nine families, are almost restricted to the northern hemisphere, although a small number belonging to the family Plethodontidae, or lungless salamanders, occur in northern and central South America. Most are terrestrial but a few live in water and others are arboreal. Apart from a few primitive species, they have internal fertilization, a feature that separates them from all but a few frogs and toads. The majority lay eggs but a few European species are viviparous, either giving birth to advanced tadpoles or, in the case of the four species: Luschan's salamander, the alpine salamander, Lanza's alpine salamander and some forms of the fire salamander, giving birth to fully developed terrestrial young.

ABOVE Salamanders occur in damp places especially in the northern hemisphere. The American Northwest where this California slender salamander, *Batrachoseps attenuatus*, lives is especially rich in species.

The frogs and toads, with 5,858 species at the time of writing, are divided into 49 families, although some authorities recognize fewer. Recent advances in molecular studies and techniques have resulted in an upheaval in the classification of frogs and toads but not all of the recently proposed changes are universally accepted (see p.133).

In fact, hardly any two experts agree. Frogs and toads are the most widespread group of amphibians and can be found on every continent except Antarctica, including many small islands. They are tail-less as adults and, with very few exceptions, fertilization is external. Frogs have filled a wide variety of ecological niches, and they include aquatic, burrowing, terrestrial and arboreal species.

ABOVE Frogs are unlikely to be mistaken for any other type of animal. This frog, a file-eared tree frog from Borneo, is an arboreal species but others live on or under the ground, or in water.

Origins of amphibians

The early ancestors of amphibians were lobe-finned fishes, Sarcopterygii, that first appeared in the Devonian Period about 380 million years ago. Lobe-finned fishes had bony connections between their fins and their vertebral column, rather than the cartilaginous structure of most other fishes' fins, paving the way for load-bearing limbs. Most of the lobe-finned fishes died out long ago but a few, including the coelacanths and the lungfish, have survived up until the present.

During the 30 million years or so since they appeared, lobe-finned fishes radiated into many distinct forms, some of which evolved structures that would be significant in the colonization of the land and the eventual arrival of amphibians. Their fins became limbs, although at first these stuck out of the sides of their bodies and were used simply to push them through the dense aquatic vegetation in which they lived and hunted. Only later did the limbs become jointed in such a way that they could make contact with the substrate allowing them to walk along the lake or seabed and, later, to support the fish's weight. Features such as this are seen in a species called *Tiktaalik*, found quite recently in Arctic Canada, which was covered in scales and whose fins ended with rays, not digits. Digits were present in other species from the same time, however. *Acanthostega* had eight digits but no wrist joints, whereas another lobe-finned fish, *Ichthyostega*, had seven digits and jointed limbs with which it could possibly have supported itself on land.

Lobe-finned fishes also had lungs, similar to those of present day lung-fish; indeed, many modern fish which evolved independently of the lobe-finned fish have auxiliary lungs and primitive lungs that were also the forerunners of swim bladders, so this was not unique to lobe-finned fishes. Other useful developments included: nostrils that were connected to the mouth cavity and thence to the lungs so they could breathe with their mouth shut, rather than having to gulp air, as the lung-fishes and other air-breathing fish do; a flexible neck that allowed the head to be turned and tilted for capturing prey; and a substantial skeleton that was capable of supporting the body when it crawled over land, thus preventing the internal organs from being crushed. All these features appear in a variety of fossil fishes at the end of the Devonian Period, about 350 million years ago, although not all necessarily in the same species.

Over time, then, these animals developed the potential to drag themselves out of the water and survive on land, at least for short periods. These early tetrapods are the ancestors of all amphibians, reptiles, birds and mammals. Then, following the early tetrapods of the late Devonian Period, about 350 million years ago, there is no trace of any intermediate fish/amphibian until the appearance of the first frog-like animal, a Pro-anuran called *Triadobatrachus*, over 100 million years later. Reptiles did not evolve from amphibians, as is often thought. Although they have common ancestors, reptiles' evolutionary pathway branched off long before modern amphibians (frogs, salamanders and caecilians) had appeared.

The early ancestors of the amphibians, the ones that bridged the gap between lobe-finned fishes and the first recognizable amphibians, have long since disappeared, leaving

BELOW The coelacanth is a surviving example of the lobe-finned fishes, the group that eventually gave rise to the amphibians.

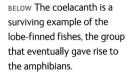

no trace. But some of their contemporaries that were evolutionary dead-ends, which arose from lobe-finned stock, are known from fossils. They included species that returned to an aquatic lifestyle (or, more likely, never completely left it) and which possessed gills as well as lungs. Remains of these early tetrapods have been found all over the globe, but they bore little resemblance to the small, inconspicuous salamanders and frogs that live in the modern world. On the contrary, they included large, predatory, crocodile-like species, many of them covered in scales. One of the largest was *Mastodonsaurus*, which lived during the Triassic Period and grew to about 4 m (13 ⅓ ft) in length with a skull measuring 125 cm (50 in), well armed with conical teeth. It was aquatic or semi-aquatic and fed primarily on fish but others became better adapted to a terrestrial life. One group of species was smaller and has only been found as larvae, perhaps because the adult forms moved away from water and lived in habitats that were not conducive to fossil-forming, or because, like some modern salamanders, they retained larval characteristics throughout their life.

It is generally agreed that all three orders of modern amphibians belong to a group known as Lissamphibia (smooth amphibians). Although there is a possibility that the caecilians are not as closely related to the salamanders and the frogs as these two groups are to each other, the general feeling is that all three orders of amphibians share so many common features – two types of skin gland (mucous and granular) and the architecture of the teeth, ear, eye and rib-cage – that they all arose from a common ancestor. Caecilians and salamanders, by the way, do not appear in the fossil record until 190 and 165 million years ago, respectively.

Origins of frogs and toads

Amphibian fossils are scarce, probably due to a combination of their relatively soft bodies and small bones (although other soft-bodied groups are well represented) and their habitats. Only animals that live in, or are washed into, mud or silt are likely to be fossilized and it may be that early amphibians lived in habitats that were unsuitable for fossilization, or it just may be that no one has found them yet. Whatever the reason, the lack of fossil frogs makes their history difficult to trace. The earliest amphibian that is identifiable as frog-like is called *Triadobatrachus massinoti*, which lived in what is now Madagascar about 230 million years ago in the Triassic Period. *Triadobatrachus* had a short tail however, and had more vertebrae than any modern frog. The next youngest is a species from northern Europe, *Czatkobatrachus*, which lived about 225 million years ago. The earliest frog fossils that can be regarded as modern are *Prosalirus bitis*, from North America and *Vieraella herbsti*, from Argentina, which are about 150–155 million years old and, a little later, a Patagonian species, *Notobatrachus degiustoi*. These frogs lacked tails, had long hind limbs and several other skeletal characteristics that identify them as relatives of the most primitive living frogs (the tailed frogs and the New Zealand frogs). Another early frog was discovered in China in 2001. It is about 125 million years old and bears some resemblance to members of the Bombinatoridae, which includes the fire-bellied toads, *Bombina*.

Diversification and classification

Throughout their collective range, frogs have spread out into a variety of habitats and adapted to the prevailing conditions there, resulting in a variety of shapes, colours and behaviours. The driving force behind these changes is natural selection. This is an ongoing process and frogs will continue to evolve and new species appear, while others die out. Nobody has yet come up with an all-encompassing definition of what a species is, but, for our purposes, we can say that frogs that are similar to each other and are able to interbreed under natural conditions, belong to the same species. Species that are similar to each other and are assumed to have arisen from a common ancestor but which, under natural conditions, do not interbreed, are placed in the same genus. Closely related genera are grouped together into families. This is a simplified view of classification. There are problems associated with it, in particular how the different levels of classification – species, genera, families, etc. – should be dealt with and there are several schools of thought. Although this need not bother us too much in an account of the natural history of frogs, a simplified overview of the current situation is given on pp.14–17.

Nomenclature

Nomenclature, the naming of species, is important, because this is how we communicate information. Common names tend to be unreliable because they may differ from place to place and because, on a global scale, there is an obvious language problem. Some species have more than one common name: the Australian species *Limnodynastes dorsalis*, is variously known as the banjo frog, sand frog, bullfrog or pobblebonk, for instance. On the other hand, some common names can apply to more than one species in different places. A 'bullfrog' is a different species depending on whether you live in North America, South Africa or Australia, for instance, and a 'green tree frog' means something different to an American, a European or an Australian. This can lead to confusion and the binomial system of scientific names, devised by Linnaeus and based on Latinized spellings is the best solution we have. It is more stable, at least at the species level, because each new name is subject to stringent review before it is accepted or changed.

BELOW *Limnodynastes dorsalis*, a frog with several common names, from Western Australia.

This binomial system of naming is illustrated in the midwife toad, *Alytes obstetricans*, and the spring peeper, *Pseudacris crucifer*. In both cases, the first word – the one given a capital letter – denotes the genus and the second name, always written with a lower case initial letter, denotes the species. Therefore, the scientific name just tells you the species and genus. What it does not do, however, is tell you which family the species belongs to; if you want to know this you have to look it up. It is also worth bearing in mind that, because the generic names reflect the relationships of each species with others, they too are subject to change if the members of a genus are re-arranged following further research. (In fact, the specific name sometimes changes too, because it has to agree with the genus, so if the generic name changes from masculine to feminine, for example, the specific name also has to change to match it.) This can be confusing and irritating to naturalists. Note also that some experts further divide species into subspecies, to denote a distinct population or colour form, but the subspecies concept is not well-defined.

What's in a name?

When it is first described, every organism is given a scientific name consisting of at least two words, the generic name and the specific name. The frog illustrated is the European tree frog and its scientific name is *Hyla arborea*. Hyla is the name of the genus and is always spelt with an uppercase initial letter, and arborea is the specific name and is always spelt with a lowercase initial letter. In this case, the generic name comes from Greek mythology, Hylas was one of the Argonauts who was abducted by wood nymphs, and the species, or specific, name describes its habitat, from the Latin *arboreus*, meaning 'of the trees'.

Species may also be named after a scientist or other person, either because they are connected with its discovery or in recognition of their contribution to herpetology. For example, one of the harlequin toads, *Atelopus boulengeri*, and 15 other frogs are named after George Albert Boulenger, a Belgian zoologist who worked at the British Museum between 1880 and 1920. Boulenger himself named several dozen new species of frogs. Others are named after their place of origin, as in the reed frog *Heterixalus madagascariensis*, or the frog *Rana cascadae*, named after the Cascade Mountains in North America, where it lives. They might be given a descriptive name, like *Boophis viridis*, where *Boophis* means 'cow-eyed' and viridis means 'green'. Nowadays, most descriptions of new species include a note on the derivation of the name (its etymology) but the meanings of some of the more obscure names given to species described long ago may have been lost.

The same specific name can be given to two or more species (many frogs have *viridis* as their specific name, for instance) and of course the same generic name is used for all the members of that genus, but no

TOP A European tree frog, *Hyla arborea*, from Italy.

ABOVE A green bright-eyed frog, *Boophis viridis*, from Andasibe, Madagascar.

two species can have the same generic and specific name, so there can only be one *Hyla arborea* and *Rana cascadae*. A small number of scientific names repeat the same word twice, as in *Bufo bufo*, the European common toad. These are known as tautonyms and this is perfectly acceptable, although it does seem to show something of a lack of imagination on the part of the name-giver – Linnaeus in this case. *Bufo* is simply the Latin word for toad.

Frogs or toads?

The difference between the term frog and toad, can be confusing. In general, we tend to refer to the smooth, slimy species as frogs and dry, warty ones as toads. Frogs jump whereas toads hop or walk. These distinctions do not stand up to scrutiny, though. Many families contain frogs as well as toads. Some species are variously known as frogs *or* toads and there is no consistency. This is simply a matter of usage – scientists do not distinguish between frogs and toads – they are all members of the Anura. The easiest way to avoid confusion is simply to call tailless amphibians in general, frogs and reserve the term toad for members of the large family Bufonidae and for species whose names are so well-established that to start calling them something different would be pedantic. So we have midwife *toads*, and painted *frogs*, for instance, even though they belong to the same family, the Alytidae.

BELOW Streamlined species with long back legs and a smooth, moist skin are popularly known as frogs whereas less agile species with warty skins are known as toads. Scientifically they are all members of the Anura. The two species illustrated here are the Asian bronzed frog, *Rana johnsi* (left), from Vietnam and the European common toad, *Bufo bufo* (right).

Recent developments in frog classification

The study of the relationships between organisms is known as systematics, which includes the way that organisms are grouped known as classification, and the way in which the groups are named known as taxonomy. Recent studies in systematics, helped by the advances in biochemistry and DNA analysis, have led to an explosion of fresh information and a better understanding of the relationships between species of frogs. Unfortunately, the traditional method used for classifying and naming organisms, known as Linnaean taxonomy, which has stood biologists in good stead for over 250 years, cannot always properly reflect the true degree of relatedness between the various species and groups of species in the light of this new information. Bear in mind that in 1735, when Linnaeus' system was published for the first time in his *Systema Naturae*, Charles Darwin's *On the Origin of Species* had not been written, nor were the principles of inheritance known. Linnaeus simply grouped species into units that looked like each other (he did not distinguish between lizards and newts, for example). Over time, the system has been pushed and pulled to cope with new discoveries but, ultimately, it is an artificial system that seeks to arrange organisms into a hierarchy that does not take into account their evolutionary history.

New methods of classification, some of which are still being developed, seek to establish a better system, which does take evolutionary history into account, grouping them into clades, which is an evolutionary branch going back to a common single ancestor and all its descendants. This is known as evolutionary taxonomy. Once established, this system should, ultimately, be able to convey more information. At present, however, Linnaean taxonomy is the only workable system by which we can name and talk about animals in a way that everyone understands. Furthermore, the species concept, which gives us the most important taxonomic unit in Linnaean taxonomy, is relatively unaffected.

When it comes to the classification of frogs, modern methods, coupled with new ways of grouping species, have resulted in a radical re-arrangement of the families. Several landmark scientific studies in the last five years have sought to re-classify frog families in a way that more accurately reflects their evolutionary history. As a result, 49 families are now recognized, at least by some authorities. At the time of writing some degree of stability appears to have been reached, although it seems likely that there will be some more changes in the future. Therefore, rather than using the traditional arrangement of families, which is already out of date, a more recent classification is used throughout and a summary of the families is given on pp.16–17, even though some of the new families will be unfamiliar to many readers. Furthermore, where there have been name changes, the older names are given in brackets throughout the book, and are also indexed. By tackling it this way, it is hoped that familiar species that may have been moved to new genera or even new families can be tracked down (see p.188 for references relating to this topic).

Chart of families

Within the families of frogs, a broad division can be made between those that arose early on and the more recent families, the neobatrachians. The early families are sometimes called the archaeobatrachians although some biologists reserve this term for ancient, extinct groups and refer to the older living groups as mesobatrachians or basal. Regardless of the terminology, it is generally agreed that these older groups cannot all trace their pedigree back to a common ancestor (they are polyphyletic), and are relics of formerly widespread families that may have evolved independently of one another, whereas the more recent families appear to have close affinities to each other and probably *do* stem from a common ancestor (they are monophyletic).

Frog families

Archaeobatrachians ('old families')	Species	Genera	Geographical range of family
Leiopelmatidae	6	2	Northwest North America and New Zealand
Pipidae	32	5	South America and Africa
Rhinophrynidae	1	1	North and Central America (Texas to Costa Rica)
Bombinatoridae	8	2	Europe, South China, Borneo, Philippines
Alytidae	12	2	Europe and North Africa
Pelobatidae	4	1	Europe, North Africa and West Asia
Megophryidae	149	10	South China and Southeast Asia
Pelodytidae	3	1	Western Europe and Southwest Asia
Scaphiopodidae	7	2	North America

Neobatrachians ('new families')	Species	Genera	Geographical range of family
Heleophrynidae	7	2	South Africa
Sooglossidae	4	2	Seychelles Islands
Nasikabatrachidae	1	1	Western Ghats of southern India
Calyptocephalellidae	4	2	Central Chile
Limnodynastidae	44	8	Australia and New Guinea
Myobatrachidae	85	13	Australia and New Guinea
Hemiphractidae	93	5	Northern South America to Panama
Ceuthomantidae	3	1	Northern South America
Brachycephalidae	44	2	South America
Craugastoridae	114	2	Southern North America to northern South America
Eleutherodactylidae	201	4	Southern North America to northern South America and the West Indies
Strabomantidae	562	17	Central and South America
Hylidae	891	44	North and South America, the West Indies, Europe, North Africa, Asia including Japan, Australia, New Guinea
Allophrynidae	1	1	Northern South America
Centrolenidae	145	11	South and Central America
Leptodactylidae	99	4	Southern North America, Central and South America and the West Indies
Ceratophryidae	86	6	South America
Cycloramphidae	101	14	South America
Leiuperidae	79	7	Central and South America

Neobatrachians ('new families')	Species	Genera	Geographical range of family
Bufonidae	550	48	North and South America, Europe, Africa (except Madagascar), Asia; widely introduced elsewhere
Hylodidae	42	3	South America
Aromobatidae	100	6	South and Central America
Dendrobatidae	174	12	South and Central America
Microhylidae	466	54	North and South America, southern Africa, Madagascar, India, Southeast Asia, New Guinea and northern Australia
Brevicipitidae	26	5	Southern Africa
Hemisotidae	9	1	Southern Africa
Arthroleptidae	139	8	Southern Africa
Hyperoliidae	208	18	Southern Africa and Madagascar
Ptychadenidae	53	3	Southern Africa
Ceratobatrachidae	84	6	Malaysian Peninsula to the Solomon Islands
Micrixalidae	11	1	India
Ranixalidae	10	1	India
Phrynobatrachidae	80	1	Southern Africa
Petropedetidae	18	2	Southern Africa
Pyxicephalidae	68	13	Southern Africa
Dicroglossidae	170	14	Africa, southern and Southeast Asia to southern China, Japan, the Philippines, New Guinea and many Pacific Islands
Mantellidae	186	12	Madagascar
Rhacophoridae	319	13	Southern Africa, Madagascar, Southeast Asia
Nyctibatrachidae	17	2	India and Sri Lanka
Ranidae	342	16	Absent only from southern South America, most of Australia and Antarctica

Note: see the following references in Further information on p.188 for references relating to the recent classification: Frost, R., 2010; Frost et al., 2006 and Vitt, L. and Caldwell, J. P., 2009.

2 Size and shape, colour and markings

THERE ARE OVER 5,600 SPECIES OF FROGS and the differences in their shape, size and colour – in effect each frog's design – are the results of natural selection. Each species has a set of characteristics that make it successful in its particular environment.

OPPOSITE Young Reinwardt's flying frogs, *Rhacophorus reinwardtii*, are unusual in being pale blue in colour. Compare this with the photograph of an adult on p.24.

Size

Frogs range in size from 10 to 300 mm (³/₅ to 12 in) in length, although most measure between 20 and 80 mm (⁴/₅ and 3 ¹/₇ in). The largest known species is the Goliath frog, *Conraua goliath*, from West Africa, which can grow to 300 mm (12 in) or more in length and to over 3 kg (6 ³/₅ lbs) in weight. Other large species include several toads, such as Blomberg's toad, *Rhaebo blombergi* (*Bufo blombergi*) and the rococo toad, *Rhinella schneideri* (*Bufo schneideri*) from South America, which can grow to 210 and 250 mm (8 ¹/₄ and 9 ⁵/₈ in) respectively, and the Asian spiny toad, *Phrynoidis aspera* (*Bufo asper*), which reaches 215 mm (8 ³/₇ in). The largest North American frog is the bullfrog, *Lithobates catesbeianus* (*Rana catesbeiana*), at 150 mm (6 in) (note this is less than half the maximum length of the Goliath frog!) and in southern Africa the African bullfrog, *Pyxicephalus adspersus*, reaches 200 mm (7 ⁷/₈ in). The largest Australian species is the southern barred frog, *Mixophyes iteratus*, which grows to 115 mm (4 ½ in), although at 150 mm (6 in) the introduced cane toad, *Rhinella marina* (*Bufo marinus*), originally from South and Central America, is larger. In Europe, the southern form of the common toad, *Bufo bufo*, is said to reach 150 mm (6 in) occasionally, but this is exceptional and the species usually only grows to a little more than half this. In all the species listed, females grow larger than males and, although there are exceptions, this is nearly always the case in frogs.

BELOW The Goliath frog, *Conraua goliath*, is the world's largest species and is hunted locally in West Africa for food.

At the other end of the scale, there are a number of frogs that measure about 10 mm (²/₅ in). The Brazilian golden toad, also known as Izecksohn's toad, *Brachycephalus didactylus*, is often cited as the smallest, at 8.6–10.4 mm (¹/₃–²/₅ in). Other miniatures, none of which have common names, include the recently described *Eleutherodactylus iberia*, from Cuba, at about the same size, and two other Cuban *Eleutherodactylus* species, *E. limbatus* and *E. orientalis*, which are marginally larger. *Noblella pygmaea*, recently described from the Peruvian Andes, has a maximum size of 12.4 mm (½ in) and the Madagascan *Stumpffia tridactyla*, and *S. pygmaea* grow to around 11 or 12 mm (*c.* ½ in). Accurate measurements of frogs as small as these are difficult and, of course, it is impossible to be sure that the specimens measured are typical as there may be larger individuals waiting to be found. Interestingly, all these small species lay their eggs on the land: the two *Stumpffia* species lay them in foam nests among leaf-litter and the tadpoles develop without feeding, and the others have direct development.

Natural selection

Populations of organisms are subject to a variety of selective pressures acting on them. These pressures include avoiding predators, finding enough food, and coping with the physical environment in which they find themselves. If they manage this long enough to reach sexual maturity, they can then attract a mate and so pass their genes on to the next generation.

Every individual living now is the descendant of a lineage that has run the gauntlet of these and other obstacles over thousands of generations, and survived. There are no unsuccessful designs – these would have been weeded out as soon as they appeared. Natural selection works because variation within a population results in some individuals being 'fitter' than others. Those traits that increase their chances of surviving and reproducing will be more likely to spread throughout the population and in this way populations change imperceptibly over many generations. As populations become fragmented and isolated, they change in different ways, and the longer they are isolated the greater the degree of change, all other things being equal.

Changes that affect the physical appearance of the organism, morphological variation, are the ones that naturalists and scientists use to identify the different species in the field. Behavioural and physiological changes are less obvious to the casual observer but are equally important.

In frogs, for instance, different populations that look alike but which have different calls avoid inter-breeding and are therefore genetically isolated from each other; they are known as cryptic species. The American leopard frog is a good example of a cryptic species in which up to 14 distinct species are recognized by some authorities because, although they all look very similar, and were formerly lumped together as one species, *Rana pipiens*, they have subtle behavioural, physiological and morphological differences.

LEFT In the past all the American leopard frogs were classified as a single species but authorities have now recognized that they are actually a group of several similar species, of which the one illustrated here is the northern leopard frog, *Lithobates pipiens*,.

Shape

Most of the more familiar frogs are generalists that can jump, swim and climb fairly well. These species have moderately slender bodies tapering towards their waist, long hind legs and shorter front ones. Their toes are usually long and lack modifications such as adhesive toe pads or hardened tubercles for digging. Webbing, which is present in many species, is restricted to the hind feet. Their eyes are large, bulbous and may be on the sides of their heads or towards the top. The eardrum, or tympanum, is just behind the eye and often about the same size.

Frogs that have more streamlined bodies, sharply pointed heads and longer hind legs, such as the Australian rocket frog, *Litoria nasuta*, the African grass frogs in the genus *Ptychadena*, and long-legged members of the Asian genus *Hylarana*, are more agile and capable of prodigious jumps of many times their own body length. These species forage around the edges of marshes and ponds and are always within one or two leaps of safety. At the other extreme, species such as the African rain frogs, *Breviceps* species, and the Australian turtle frog, *Myobatrachus gouldii*, are almost spherical in shape with short limbs. These very rotund frogs cannot leap at all and are burrowing species that live underground for much of the time, and even breed in subterranean chambers. They feed on ants and termites, and eat huge numbers at a sitting, requiring little effort to capture them. Agility is not a high priority.

A distinct group of frogs that have stouter than average shape and short limbs are predatory species such as the South American horned frogs, *Ceratophrys* species, and the African bullfrogs, *Pyxicephalus adspersus* and *P. edulis*. These species do not chase after their prey but wait for it to wander within range, and then pounce on it. Their mouths are enormously wide, necessitating a body shape to match.

LEFT The rocket frog, *Litoria nasuta*, is a good example of an agile long-limbed, streamlined species.

Limbs

The shape of the limbs and feet are of great significance because they often provide clues to the life-style of the frogs concerned. Long-limbed species may be agile terrestrial leapers, such as the species mentioned above, or they may be arboreal species, where the greater reach and leverage of lengthy limbs gives then an advantage. In addition, climbing species nearly always have digits that end in terminal discs. The adhesive properties of these discs, or toe pads, are vital when climbing in trees, shrubs or on rock faces.

The toe pads have to grip smooth as well as rough surfaces and, at the microscopic level, their structure is quite complicated. They are covered with numerous interlocking, flat-topped columns that interlock leaving a network of narrow spaces between them. Glands secrete mucus into these spaces and the mucus moves along them by capillary action. On rough surfaces, such as bark, the columns act like bristles to provide grip through friction. On smooth surfaces, such as shiny leaves, the channels of mucus create adhesion through the force of surface tension. Although they are sometimes called suctorial pads, suction does not play a part in their function; if the surface they are clinging to is too wet, they are unable to grip.

ABOVE Like most members of the Rhacophoridae, Fea's treefrog, *Rhacophorus feae*, from Vietnam uses expanded toe pads to help it grip vertical surfaces.

BELOW A scanning electron microscope image of the surface of the toe pad of a climbing frog.

Adhesive toes are common to members of many families not just the tree frogs of the Hylidae. Among others they also occur widely in the Rhacophoridae, Hyperoliidae, Centrolenidae, Ranidae, Mantellidae and Microhylidae, for example. The pads are of varying shapes and sizes and the underlying structure varies between the families, indicating that this is a characteristic that has evolved more than once among frogs. For example, tree frogs belonging to the Hylidae have a small piece of cartilage between the last two bones of the toes (the intercalary cartilage) that probably increases the flexibility of their tips, and a groove runs around the margin of the pad, delineating the upper surface from the adhesive section. Dendrobatids, however, have a pair of fleshy flaps on the tips of their digits and lack the intercalary cartilage. Treehole frogs, *Metaphrynella* spp., which belong to the Microhylidae, have elongated adhesive pads on the upper sections of their fingers as well as those on the tips of their digits.

A hardened spade, in reality a sharp-edged tubercle, often grows on the inside edge of the hind feet of frogs that burrow, including the European and American spadefoot toads, *Pelobates*, *Scaphiopus* and *Spea* and the remarkably similar *Heleioporus* species from Australia. A number of burrowing toads belonging to the Bufonidae also have spades. This structure is often coupled with short hind limbs, short toes and muscles that help to rotate the frog's foot as it digs so that the spade is used to good advantage.

All frogs that swim have powerful hind limbs and webbing between their toes but highly aquatic species are especially well adapted, with long digits and webbing that reaches almost to their tips. The clawed frogs and pipa toads, Andean water frogs, *Telmatobius* spp., the rare *Barbourula* spp. and, to a lesser extent, their relatives the fire-bellied toads, *Bombina* spp., fit this description.

The front limbs of aquatic species often lack webbing and are not much used for swimming. For example, the pipids and the fire-bellied toads use their front limbs to stuff prey into their mouths. For frogs that enter the water only occasionally, or only during the breeding season, there must be a trade-off between the ability to swim well, on the one hand, and to move about on land on the other and so they often have less extensive webbing.

Two small groups of species, popularly known as flying frogs, have heavily webbed front feet as well as back. These frogs do not enter the water but spread their digits to allow them to parachute gently from tall trees. By altering the angle of their feet they can, to some extent, control their glide, which can be at an angle of up to 45° in some species, and they can also turn. This technique is used primarily to escape from predators although it may also give them a quick route to the forest floor during the breeding season.

BELOW Spadefoot toads of many kinds have a hardened, crescent-shaped tubercle on their hind feet for digging. This is a western spadefoot toad, *Pelobates cultripes*, from Spain.

Convergent evolution

Convergent evolution is the result of unrelated species finding similar solutions to similar problems so that they look and behave like each other. Arboreal frogs with expanded toe pads are one example, and the digging spade on the hind feet of many unrelated burrowing frogs is another, but perhaps the best example is the case of the so-called flying frogs, of which there are two distinct groups. In New World rainforests, tree frogs of the genera *Agalychnis* and *Phyllomedusa*, have huge, heavily webbed feet that they use to glide from the forest canopy to lower levels. In Southeast Asian rainforests, an unrelated group of frogs, belonging to the genus *Rhacophorus*, do exactly the same thing and have evolved along similar lines. Differences between, say, Spurrell's flying frog, *Agalychnis spurrelli*, from Central America and Reinwardt's flying frog, *Rhacophorus reinwardtii*, from Malaysia and Indonesia, are so slight that they could easily be mistaken for each other at first glance, even though they belong to two different families: the Hylidae and the Rhacophoridae. Incidentally, none of these frogs breed in open water but instead lay egg-masses on leaves overhanging pools. They rarely if ever enter the water – ironically, the extensive webbing that allows them to parachute to the ground is not suitable for swimming.

ABOVE Spurrell's flying frog, *Agalychnis spurrelli*, a member of the Hylidae from Costa Rica.

BELOW Reinwardt's flying frog, *Rhacophorus reinwardtii*, a member of the Rhacophoridae from Borneo.

Eyes

Frogs' eyes may be large or small, depending on how much reliance they place on vision. Those of active, nocturnal species are among the largest, whereas subterranean species (and aquatic frogs that live in murky water) may have very small eyes. The shape of the pupil also varies. Nocturnal frogs have pupils that close down to narrow slits in bright light. The pupils

may be horizontal or vertical and, although the shape of the pupils tends to be constant among closely related frogs, both types can be found in some families, such as the Hylidae and the Hyperoliidae. The Microhylidae, which includes a varied assemblage of frogs, includes species with horizontal, vertical and round pupils. Fire-bellied toads, *Bombina* spp., have pupils that are shaped like inverted tear-drops and those of puddle frogs, of the genus *Occidozyga*, are of a similar shape; both groups of species live in shallow pools and habitually rest with just their eyes above the surface. Madagascan reed frogs, *Heterixalus* spp., have vertical pupils that are rounded on one edge and slightly angular on the other.

ABOVE AND LEFT Pupil shape can vary even within the same family. This red-footed tree frog, *Hypsiboas rufitelus*, from Costa Rica has horizontal pupils, whereas those of the lemur leaf frog, *Phyllomedusa lemur*, also from Costa Rica are vertical. Both are members of the Hylidae.

OPPOSITE LEFT The green bright-eyed frog, *Boophis viridis*, from Madagascar has unusual blue rims to its irises.

The colour of the iris often matches that of the frog, and the iris may have a horizontal dark line passing through to disguise it. Some species that are otherwise dull can have surprisingly beautiful eyes. The European common toad has a bright coppery coloured iris, reticulated with fine black lines. Even Shakespeare was impressed: 'Sweet are the uses of adversity, Which, like the toad, ugly and venomous, Wears yet a precious jewel in his head.' Others are green or yellow and the bright-eyed frog, *Boophis viridis*, from Madagascar has a blue rim to its irises. Red is the dominant colour of the irises of the aptly named Central American red-eyed leaf frog, *Agalychnis callidryas*, the Madagascan red-eyed tree frog, *Boophis luteus*, and the Australian red-eyed tree frog, *Litoria chloris*. The red is due to the pigment rhodopsin, which is especially sensitive to light in the green-yellow range, and may help the frog to see better as fading light is filtered by a canopy of leaves in the evening and early part of the night, when the frogs are most active.

RIGHT Like many bufonids, the European common toad, *Bufo bufo*, has beautiful copper-coloured irises.

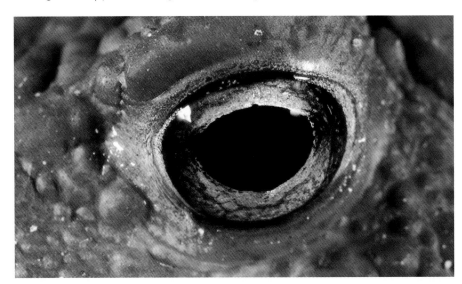

RIGHT The aptly-named red-eyed leaf frog, *Agalychnis callidryas*, from Central America has startlingly bright red irises due to pigments that are thought to improve its eyesight in the dim light beneath the forest canopy.

The eyes are protected by moveable upper and lower eyelids. There is also a third eyelid, the nictitating membrane, which is positioned under the lower eyelid when at rest. It is usually transparent and the frog closes it when swimming, and also uses it to wipe the surface of its eye. In some species, especially certain tree frogs like the red-eyed tree frog, *Agalychnis callidryas*, the geographic tree frog, *Hypsiboas geographicus* (*Hyla geographica*), and the Australian lace-lid, *Nyctimystes dayi*, the nictitating membrane is semi-transparent and patterned with intricate reticulations. The frogs close it when they are resting during the day so that they can continue to monitor their surroundings while the surface of their eyes is protected (and perhaps also less conspicuous).

TOP Camouflaged frogs often have eyes that match their coloration, none more so than the mossy frog, *Theloderma corticale*, from Vietnam.

ABOVE The nictitating membrane of some frogs, such as the red-eyed leaf frog, is intricately marked.

Ears

Most frogs vocalize in order to attract mates and defend territories and so it is essential that they can hear, too. Their ears are beneath the surface just behind their eyes and their presence is marked in most species by a circular area of skin known as the tympanum. This varies in size from species to species and in some frogs, between males and females. In American bullfrogs, for instance, males' tympani are about twice the diameter of females' tympani.

Frills and flaps

Many frogs, such as the frilled tree frog, *Rhacophorus appendiculatus*, have flaps of skin fringing their bodies and limbs. A common feature is the small triangular flap of skin on the heels and sometimes also on the elbows of a number of unrelated tree frogs. Others have small 'horns' above their eyes. All these help the frog to break up its outline and escape the notice of predators. Males of the hairy frog, *Trichobatrachus robustus*, develop rows of thread-like papillae, which are extensions of their skin, along their backs and thighs in the breeding season. They increase the skin's surface area allowing the frogs to increase their oxygen uptake from the fast-running streams in which they breed. Folds and ridges in the skin of other species, such as many ranids, increase its surface area and probably help in respiration and water balance. The warty or tuberculate skins of toads and other frogs owe their texture to concentrations of the granular glands that secrete various substances to keep the frog free from bacterial and fungal infections, and deter predators. These tubercles may be scattered randomly over the dorsal surface and limbs or they may be concentrated in certain areas, especially behind the head. In some species, such as the mossy frog, *Theloderma corticale*, the purpose of these tubercles may have more to do with disguising the frog's appearance. There is more on these topics in chapters 3 and 4.

OPPOSITE TOP The tympanum, is not always visible but is obvious behind the eye of this northern leopard frog, *Lithobates pipiens*, from North America .

OPPOSITE BOTTOM Male bullfrogs, *Lithobates catesbeiana*, have larger tympanums than females. As well as a sound-receiving organ, the large tympanum of this species vibrates when it calls, helping to broadcast the sound from both sides of its head.

BELOW Horned frogs, *Ceratophrys* spp., owe their common name to the small pointed extensions of skin above their eyes.

Colour

Shades of brown and grey are by far the most common colours among frogs. This makes sense considering that most live on the ground, among short vegetation or in leaf-litter. The next most common colour is green, found in arboreal species of many types, and some terrestrial ones. Only a small proportion of frogs are brightly coloured and these are usually species that use colour, in conjunction with black, to advertise the fact that they are toxic. The poison dart frogs and the mantellas are excellent examples of this and, between them they display the full spectrum of colours, from bright red to bright blue. A few frogs have patches of brightly coloured skin on parts of the body that are normally hidden, such as the groin. This is known as flash coloration and is discussed in Chapter 4. Most of the colours are produced by pigments in the skin. These pigments are contained in specialized cells known as chromatophores. There are four types of chromatophores as follows.

Melanophores contain melanins, which produce black, brown or red colours. The melanophores are the most important colour-producing cells and are present in large numbers in the exposed surfaces of the skin. Stripes, spots and blotches are caused by the distribution pattern of melanophores over various parts of the frog's surface. The pigment granules they contain can be spread throughout the cell or they may clump together to form a small speck in the centre, and how they are distributed is under hormonal control. Generally speaking, cold, damp conditions cause the pigment granules to spread out, so the frog becomes darker overall whereas warm, dry conditions cause the granules to contract, and the frog becomes lighter in colour. Changes can happen quickly, sometimes in a matter of minutes.

RIGHT Poison dart frogs from the tropical forests of Central and South America obtain their coloration from pigments in their skin. In the harlequin poison dart frog, *Oophaga histrionica*, from Ecuador, the predominant pigment is red.

LEFT Another poison dart frog, *Dendrobates tinctorius azureus*, is remarkable for its bright blue coloration. The purpose of the bright colours is to warn predators that their skin contains dangerous toxins.

Xanthophores and erythrophores contain carotenoids and pterines, which produce yellow, orange or red colours. The pigment in these chromatophores is not as mobile as that of the melanophores and so the effects they produce are not as changeable, at least in the short term. The substances that make them are derived from the frog's food and they vary in intensity according to its diet. Carotenoids are present in the leaves, stems and tubers of many plants, and frogs obtain carotenoids by feeding on plant-eating insects, for example. Individuals of normally, brightly coloured species that are deprived of a natural diet, in captivity for example, often lack the bright colours that their wild counterparts display. Captive-raised fire-bellied toads for example have yellowish instead of red bellies due to the absence of carotenoids in their diet.

Iridophores contain plate-like structures made primarily of a white substance, guanine. They are located deeper in the skin than the melanophores, erythrophores or xanthophores. They contain structures that absorb long wavelengths, those at the red end of the spectrum, and reflect back only the rays from the blue end of the spectrum. This is known as Tyndall scattering and is the same effect that gives the sky its blue colour, which is created by reflections from small particles in the atmosphere. In frogs, the blue light reflected back from the iridophores usually passes through a filtering layer of yellow xanthophores, giving the frog its green colour. If the xanthophores are absent, through a mutation, a normally green frog appears blue; if the iridophores are absent, it is yellow. Both forms have been recorded in the stripeless tree frog, *Hyla meridionalis*. Captive frogs that should be green are often dull bluish-grey because the amount of pigment in the xanthophores is insufficient to filter the blue light effectively. Where all pigments are absent, due to a breakdown in the biochemical pathway that creates pigments caused by a mutation, the resulting frog is an albino.

Colour changes

Apart from the ability to become darker or lighter in response to temperature and humidity, there are other examples of colour change. It is quite common for frogs that are dark by day to become paler at night. The back and exposed limb surfaces of the spotted tree frog, *Hypsiboas punctatus* (*Hyla punctata*), change from bright yellowish green in the day to dark red at night. Some frogs change colour during the breeding season. Males of the moor frog, *Rana arvalis*, turn blue or purple in some parts of their range, while male Madagascar jumping frogs, *Aglyptodactylus madagascariensis*, Costa Rican meadow frogs, *Isthmohyla pseudopuma* (*Hyla pseudopuma*), lesser tree frogs, *Dendropsophus minutus* (*Hyla minuta*), and several small Australian *Litoria* species, turn yellow. Why this should happen is a mystery.

A few change colour as they grow. Mantellas, for example are dull brown when first they metamorphose but become brighter as they grow, perhaps because they are not armed with toxins at first. Poison dart frogs, by contrast, tend to be brightly coloured from the word go, although the markings of juveniles may be less intricate than those of adults. Many green frogs start life as brown juveniles and their adult coloration develops as they grow, and Reinwardt's flying frogs, *Rhacophorus reinwardtii*, are pale blue, almost white, when they metamorphose but bright green as adults.

Colour variation

Populations of some frogs include individuals with two or more distinct colours or markings. This is known as polymorphy. Polymorphic populations gain protection from predators that are unable to build up a search image of every form; there is more on this in Chapter 4. A common type of polymorphy in frogs occurs when some individuals have a pale stripe down the centre of their back while others do not. In other examples, individuals may be plain, striped or spotted, or they may be coloured differently, as in the Madagascan reed frog, *Heterixalus madagascariensis*, of which there are several colour forms.

Sexual dimorphism

Sexual dimorphism where males and females have different colours, is unusual among frogs. Because males attract mates by calling, usually at night, there is little need or point, in being brightly coloured, especially as this may attract predators during the day when they are resting. Leaving aside the examples given above in which males change colour temporarily during the breeding season, there are just a few examples. The most spectacular is the Monte Verde toad, *Incilius periglenes* (*Bufo periglenes*), now sadly extinct, in which males were bright orange and females were red with large brown blotches. Females of the Yosemite toad, *Anaxyrus canorus* (*Bufo canorus*), are heavily spotted whereas males are plain brown or lightly speckled and a similar situation exists in Couch's spadefoot toad, *Scaphiopus couchii*. In the argus reed frog, *Hyperolius argus*, males are pale green in colour whereas females are dark brown with cream spots and stripes, and in *Hyperolius riggenbachi*, males are also pale green or olive but females are black with cream squiggles, or they have a complicated pattern of red, black and yellow vermiculations, depending on where they live. The Asian tree toad, *Pedostibes hosii*, has an interesting variation on this theme: all the males are brown and about half the females are also brown but the rest of the females are olive to dark purple with bright orange spots.

LEFT AND BELOW Most male argus reed frogs, *Hyperolius argus* (left), are green whereas most females (below) are brown with yellow spots, a rare example of such pronounced sexual dimorphism.

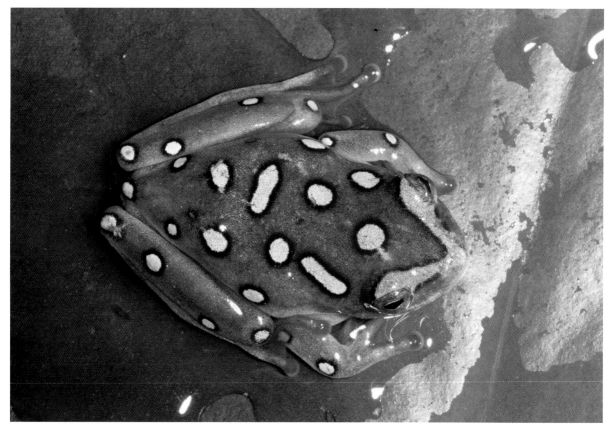

ABOVE A pair of Asian tree toads, *Pedostibes hosii*, in axillary amplexus. This species demonstrates sexual dimorphism with all the males being a brown colour and about half the females an olive colour with yellow spots. The other half of the females are brown like the males.

Identification

Identifying frogs and toads is the starting point for further study and research. In any case, it is always satisfying to put a name to something you have seen – ask any birdwatcher. How hard this will be depends on where you are: in the United Kingdom, where we only have three native species, it is easy but in other parts, especially the tropics, it can be next to impossible and even experts can get confused. Characteristics that separate one species from another are those discussed previously: size; general body shape; pupil shape; the amount of webbing on the back and front feet; any obvious glands or ridges on the skin's surface; colour and markings (with some caution as many species are variable). All this may seem pretty obvious but it is annoying to get back home to find that one species can only be separated from another by some feature that you did not look at. If you get a chance to take a photograph this might also be helpful. Handling or otherwise disturbing some rare species may be illegal. Similarly, frogs in national parks or other protected areas should not be disturbed. If you do catch a frog to confirm its identity, or to photograph it, release it as soon as possible in the same place that you caught it. Breeding frogs can often be identified from their calls, even when they cannot be seen, and recordings are available for some areas, or are included with field guides.

Field guides are available for many of most-visited regions (see p.188), but there are large gaps – West Africa, much of South America, parts of Southeast Asia – where literature is scarce or non-existent. Other identification guides are so bulky that they can hardly be called field guides but you can use them to research an area you are visiting for the first time before you leave home. The internet can also help but should be used with caution – only reputable sites, especially those maintained by universities and other educational organizations, can be safely relied upon (see p.188).

BELOW The colours and markings of frogs and toads often vary, sometimes making identification difficult. Some grass frogs like *Fejervarya limnocharis*, have a stripe down their back whereas others do not . Variations like this are found in several other species.

3 Interactions with the physical environment

RESPIRATION, WATER BALANCE AND THERMOREGULATION are the most important ways in which amphibians interact with their physical environment. These three functions are all interconnected, and the skin plays a vital role in maintaining a balance between them, and it also gives some protection against physical damage.

The skin

The skin consists of two layers, the dermis, or inner layer, and the epidermis, or outer layer. Frogs shed the outermost part of their epidermis, known as the stratum corneum, at intervals that can be as short as a few days, often contorting their body and limbs to work it free and using their mouth to pull it off before eating it.

Chromatophores, the cells that give the frog its colour and pattern, described in Chapter 2, are located in both the dermis and the epidermis. In addition, there are two or more types of glands embedded in the dermis and connected to the surface via pores in the epidermis. The granular glands, or poison glands, produce substances that guard the skin against bacterial and fungal infection and, in many species, also provide the frogs with some protection from larger predators by being distasteful or toxic. Poison glands are more numerous on the frog's back, head and, to a lesser extent, the outer surfaces of

OPPOSITE Reed frogs related to *Hyperolius viridiflavus* from East Africa have skin that reflects light and heat, allowing them to rest in exposed positions during the heat of the day.

BELOW LEFT Bufonid toads and many other species have enlarged warts in their skin that indicate places where there are concentrations of poison glands. The glands are especially numerous on the dorsal surface of marine or cane toads, *Rhinella marina*.

BELOW RIGHT Frogs' skin contains mucous glands to keep the skin moist, which is essential for cutaneous respiration, and chromatophores, which are the cells that produce colour and pattern.

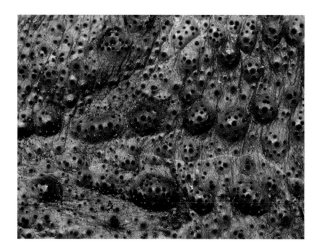

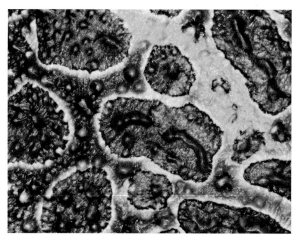

its limbs. They are often grouped into aggregations forming the enlarged 'warts' of toads and other species. The fluid they secrete is usually milky white or yellowish and may be sticky and it varies in its potency, being highly developed in species such as the poison dart frogs, Dendrobatidae, and in some toads.

The mucous glands secrete fluids that keep the skin moist, allowing gaseous exchange – respiration – to occur across the skin's surface. The wet surface also reduces friction while they are swimming and makes them difficult for predators, including humans, to grasp. When fluid on the skin's surface evaporates it reduces the temperature of the frog and so it can also be important in thermoregulation. As the frog warms up it secretes more fluid to increase the rate of evaporation and, therefore, the rate of cooling. This can lead to dehydration however, unless the frog is able to replace the fluid lost.

In addition to granular and mucous glands, certain species of frogs have more specialized glands. Arboreal frogs, for example, have concentrations of mucous glands on their toe pads to give them their stickiness and a few species that live in very dry places have glands that produce a waxy substance, which they can smear over their bodies to protect against dehydration. Nuptial glands, found only in males, are associated with areas of rough, usually pigmented, skin on the hands, limbs and occasionally chests, depending on the species. These only appear during the breeding season and help the male to grasp slippery females during the mating process, a position known as amplexus, and they tend to be restricted to species that breed in water. In some very rotund species, such as the African rain frogs, *Breviceps* spp., in which the forelimbs are too short to reach around the female's body, nuptial glands on the male's underside secrete a sticky mucous so that he can attach himself to females during egg-laying.

Respiration

In the early stages of their development, tadpoles breathe through paired gills, like fish. Each gill is divided into at least three branches with many feathery filaments. These are well-supplied with small blood vessels so that gaseous exchange can take place. The gills are visible externally in newly hatched tadpoles but are soon covered with flaps of skin as the tadpole develops. The tadpole gulps water regularly and pumps it into the gill chambers, over the gills, and out again through an opening called the spiracle. In most tadpoles both gill chambers are connected and the water exits through a single spiracle, but some primitive species have two spiracles, one on each side. The gills are replaced by lungs at about the same time as the tail is absorbed. Aquatic frogs also extract oxygen from the water, through their skin, but must surface to take air into their lungs. How often they do this depends on several factors, the most important of which is the oxygen capacity of the water – cool running water contains much more dissolved oxygen than warm stagnant water.

Terrestrial frogs breathe through lungs and through their skin. The lungs are paired and fundamentally similar to those of other vertebrates, including mammals. Amphibian

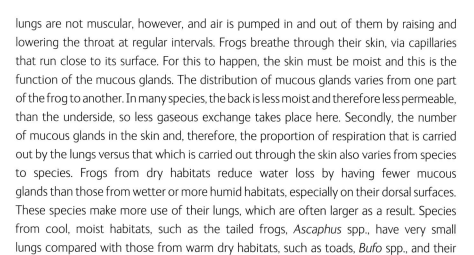

lungs are not muscular, however, and air is pumped in and out of them by raising and lowering the throat at regular intervals. Frogs breathe through their skin, via capillaries that run close to its surface. For this to happen, the skin must be moist and this is the function of the mucous glands. The distribution of mucous glands varies from one part of the frog to another. In many species, the back is less moist and therefore less permeable, than the underside, so less gaseous exchange takes place here. Secondly, the number of mucous glands in the skin and, therefore, the proportion of respiration that is carried out by the lungs versus that which is carried out through the skin also varies from species to species. Frogs from dry habitats reduce water loss by having fewer mucous glands than those from wetter or more humid habitats, especially on their dorsal surfaces. These species make more use of their lungs, which are often larger as a result. Species from cool, moist habitats, such as the tailed frogs, *Ascaphus* spp., have very small lungs compared with those from warm dry habitats, such as toads, *Bufo* spp., and their

BELOW The folded skin of the aquatic Lake Titicaca frog, *Telmatobius culeus*, provides a greater surface area over which dissolved oxygen can be absorbed.

relatives (although in this case smaller lungs also reduce the frog's buoyancy, allowing it to move about on the bottom of the stream without floating towards the surface). Superimposed on this is the activity of the frog and the temperature. As the temperature goes up and the frog becomes more active, it will need its lungs to work harder, whereas oxygen uptake across its skin will increase only slightly if at all.

Another factor is the ratio of surface to body mass so the greater the relative surface area of skin, the more oxygen it can absorb. Some frogs exploit this in unusual ways. The Lake Titicaca frog, *Telmatobius culeus*, is totally aquatic and obtains most of its oxygen through its skin. It can do this because its skin is

folded and wrinkled to such an extent that it looks as though it has borrowed it from a much bigger frog. The folds increase the skin area greatly, compensating for the fact that, at high altitude, the air and water contain less oxygen than at lower altitudes. And in the hairy frog, *Trichobatrachus robustus*, from West Africa, males develop rows of thread-like extensions of their skin during the breeding season. This species is a stream-breeder and the extra surface area may enable the male to stay underwater longer, presumably to allow him to remain in amplexus while the female attaches her eggs to the streambed. Whereas the hairy frog relies on moving water to bring fresh oxygen supplies to his skin, the Lake Titicaca frog lives in still water and so it must constantly keep its skin moving by gently swaying about, to prevent a layer of oxygen-depleted water forming around it.

Water balance

Frogs do not drink but absorb water from their surroundings through their skin. As discussed previously, they also take in oxygen and expel carbon dioxide across the surface of their skin, but this can only occur if the skin is moist. This means that the frog must keep a balance between the amount of water lost through evaporation and the amount absorbed by contact with water or a moist surface. Aquatic frogs, tadpoles and all frogs when they are in the water, even if only temporarily, have a fairly simple problem to solve: how to limit the amount of water retained in their system, and they do this by excreting large quantities of dilute urine. The task for terrestrial frogs, on the other hand, is more complicated. They have to limit the amount of water passing out of their body in order to avoid dehydration. They do this in three ways: by their behaviour, by the structure of the skin and by physiological processes.

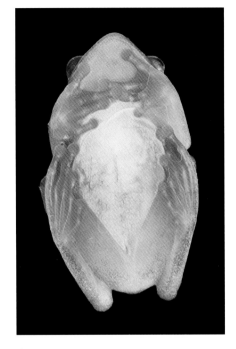

ABOVE When resting, the argus reed frog, *Hyperolius argus*, and most other frogs tuck their limbs under their bodies to reduce evaporation.

Frogs living in dry places, or places that become dry from time to time, have to behave in ways that limit water loss. They can do this by being active at night, by crawling under logs, rocks and leaf-litter by day, and by burrowing down into moister layers of soil during dry weather – this is the first line of defence against desiccation. In addition, species from drier places have skin that is less permeable, especially on the back and the exposed parts of the limbs, than those from moist habitats. When the air is dry they sit in a hunched position with their limbs pulled under their bodies to reduce the exposure of the more permeable parts. They keep the underside of their body and limbs in contact with the substrate to reduce water loss and, if possible, to soak up any moisture present. The wrinkled surface of frog and toad skin also acts like a wick to draw up moisture from wet surfaces by capillary action, so just by sitting on a damp surface the frog can blot up water quickly and efficiently.

RIGHT At the beginning of the dry season Budgett's frog, *Lepidobatrachus laevis*, from the Chaco region of Argentina burrows down into the mud and forms a cocoon around its body to avoid desiccation.

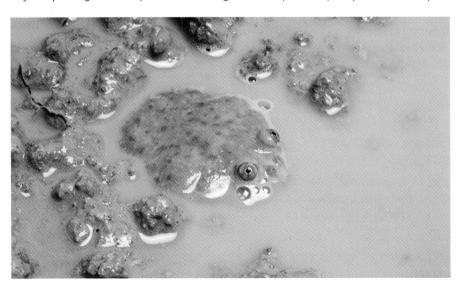

LEFT Couch's spadefoot toad, *Scaphiopus couchii*, from the Sonoran Desert of Arizona, North America, spends much of the year underground to avoid dry weather.

Some species from seasonally very dry places burrow down into the soil or mud and form a cocoon around their bodies by retaining several layers of shed skin that becomes cornified, or papery, and impervious to water, thus greatly reducing the amount of water lost by evaporation. At the same time, they lower their metabolic rate and store water in their bladder for re-absorption as required. Examples include Budgett's frog, *Lepidobatrachus laevis*, several Australian *Litoria* species (formerly *Cyclorana* species), the African bullfrog, *Pyxicephalus adspersus*, and the Mexican hylid, *Smilisca fodiens* (*Pternohyla fodiens*). Frogs can store 20–50% of their body weight in bladder water, and the bladder of the Australian water-holding frog, *Cyclorana platycephala*, can hold as much as 130% of the frog's weight. This species has even been used as a source of drinking water by native peoples.

BELOW Australian frogs of the genus *Cyclorana* are known as water-holding frogs. They can store large quantities of water in their bladders in order to survive periods of drought.

Physiologically, frogs can increase the amount of urea they produce as a waste product. Higher concentrations of urea in the tissues reduce the rate at which water is lost to the environment by osmosis, because water travels from lower concentrations to higher ones across a permeable surface. For the same reason, frogs can increase the rate at which water is taken up if the opportunity arises, so they can rehydrate rapidly. Arboreal frogs face a particular problem, even if they live in relatively moist environments, because the flow of air around their entire bodies has a drying effect. Many tree frogs have mechanisms that help to reduce water loss and most simply have a more waterproof, that is a less permeable skin than those living in contact with the ground. Some, such as the three species of African grey tree frogs belonging to the genus *Chiromantis*, not only have skin that is very waterproof, but are also able to convert their waste products to uric acid, a white paste that needs almost no water to carry it out of the body. (Birds and reptiles do the same thing.) The waxy tree frog, *Phyllomedusa sauvagii*, and at least three other related species, have specialized lipid glands in their skin that secrete a waxy substance that they wipe over their entire surface using their hands and feet. These species frequently bask in open sun in some of the hottest and driest parts of southern South America. The Asian tree frog, *Polypedates maculatus*, has similar glands and behaviour although water balance in this species has not been fully investigated. Experiments have shown that some African reed frogs, such as the marbled reed frog, *Hyperolius marmoratus*, are also resistant to drying out, far more than other tree frogs of similar size, for reasons that are not fully understood. It is not unusual to find these reed frogs resting on exposed leaves in full sun during the middle of the day.

RIGHT The Pacific tree frog or Pacific chorus frog, *Pseudacris regilla*, often rests fully exposed to the midday sun but maintains a hunched position to avoid the drying effects of the warm air.

RIGHT Waxy frogs, *Phyllomedusa sauvagii*, waterproof their skin by wiping a waxy substance over their entire bodies.

Thermoregulation

Amphibians are ectotherms, they do not produce their own body heat internally in the same way as birds and mammals but rely on outside sources to provide it. Their bodily functions such as movement, digestion, spermatogenesis, etc., will only take place within certain upper and lower temperature limits. Below these limits they slow down and eventually stop altogether. They lose the power of locomotion and, if the temperature continues to fall, they may die. Similarly, if they allow their bodies to get too hot, they die of heat exhaustion. Somewhere between the upper and lower limits each species has an optimal body temperature which it strives to attain. This is usually nearer the upper limit than the lower one and is about 25–35°C (77–95°F) for many frogs. Preferred body temperatures vary, to some extent, according to where the species is from. Those from cooler climates have adapted to a lower preferred temperature, and even where a single species has a wide north–south range, different populations may have different preferred temperatures depending on which part of the range they come from. Having said all this, many species operate far below their ideal temperature because of various constraints – if they are nocturnal, for instance, or if they live in an especially cold environment. We know this because when frogs are given a choice of temperatures under laboratory conditions they frequently choose temperatures significantly higher than those at which they are found to have in the wild. Even in tropical forests, temperatures at ground level are often lower than the preferred temperatures of the frogs that live there.

The various ways in which they regulate their temperature depend very much on the species and its preferred habitat. Behavioural temperature control includes moving from cool to warm places and vice versa, and of adjusting the position of their body so that the right amount is exposed to the sun assuming there is any. Not all species are able to thermoregulate by shuttling from place to place, or by altering their posture. If they live in a habitat that has an even temperature, such as a pond or lake, beneath the ground or among the leaf-litter of a forest floor, there is little opportunity to move somewhere warmer – they are passive thermoregulators, accepting the ambient temperature of their environment. This is fine, of course, if they happen to live in a part of the world where ambient temperatures are acceptable most of the time and is exactly why totally aquatic frogs, and most leaf-litter dwellers, tend to have tropical distributions.

Frogs living at high altitudes and in other cool environments can sometimes raise their body temperatures considerably above those of their surroundings. Dark colours absorb more heat than light ones, so frogs that are cold, or are warming up, tend to be darker than those that have reached a comfortable temperature or which are trying to cool down. (Colour change in frogs is discussed in Chapter 2.) Basking frogs, however, run the risk of dehydration. Some get around this by basking in shallow water or by resting with their undersides, where their skin is most permeable, pressed against a damp substrate to counteract evaporation – they absorb water as quickly as they lose it. As they approach their preferred temperature they may raise themselves up to allow evaporation to take place and this has a cooling effect. In this way they can achieve a balance between temperature and water content. If, on the other hand, they are not in contact with water

ABOVE Frogs that live on the forest floor such as the Madagascan jumping frog, *Aglyptodactylus madagascariensis*, have little opportunity to regulate their body temperatures and so they tend to be more numerous in places where the ambient temperature is suitable for them.

or a moist surface they will begin to exhaust their water supply after a while. The point at which this stage is reached depends on the species and its tolerance to dehydration. Tree frogs are often able to bask for longer than terrestrial ones because they usually have more protection against desiccation, as mentioned above. The body temperatures of the grey tree frogs and waxy frogs, for instance, can reach 40°C (104°F) in dry conditions without coming to harm. Another species, the Kenyan reed frog, *Hyperolius viridiflavus* (and probably some of the other reed frogs) can rest in the open during the day without overheating or dehydrating. During the dry season the reed frog converts its nitrogenous waste to guanine, a white compound also found in the droppings of birds and reptiles (and comes from the word guano). The guanine is moved to the light-reflecting cells in the skin – the iridophores – and the frog benefits twice: the amount of water required to carry its waste from its body is reduced, so helping to avoid desiccation and the reflectance of the skin is increased, keeping its temperature under control.

Why would all these frogs go to such lengths so that they can sit out in the sun when they could just as easily creep under a log or stone and conserve water? Because by raising their body temperatures they speed up their metabolism. This allows them to digest and assimilate food more quickly so they grow faster, and may help them to improve their 'fitness' in other ways as it gives them an edge over frogs that have to operate at temperatures below their most efficient. The reed frogs mentioned above grow up and breed within six months of hatching. Newly metamorphosed froglets and toadlets often leave the water during the day, presumably because the warmer conditions allow them to be more active. Increasing their metabolism more than compensates for the additional risk of predation.

Freezing frogs

Although frogs are more numerous in warmer regions, especially the tropics, there are still plenty of species in more temperate climates. Two of them, the American wood frog, *Lithobates sylvaticus* (*Rana sylvatica*), and the Siberian wood frog, *Rana amurensis*, occur above the Arctic Circle. Others are found in northern Europe and Canada, and southern South America, where conditions are such that they must hibernate for at least six months of every year by retreating to places where they are protected from predators and from the worst of the weather.

Some frogs live in places where they simply cannot get below the frost level and where the ground temperature may drop as low as minus 5°C (23°F). (A salamander from the Siberian steppes, *Salamandrella keyserlingii*, can survive temperatures as low as minus 40°C (104°F) many metres down in the permafrost.)

Species such as the wood frog, grey tree frogs, *Hyla versicolor* and *H. chrysoscelis*, and the spring peeper, *Pseudacris crucifer*, in North America, and the common frog in Europe, can survive freezing conditions as long as ice crystals do not form inside their cells. They prevent this by breaking down the glycogen stored in their liver into glucose or glycerol (or both) which they release into their bloodstream at up to 60 times its normal level. This acts as an anti-freeze, lowering the freezing point of the fluid inside the cells. Water between the cells may freeze though, and the frogs become literally frozen stiff with their metabolism shut down completely. European marsh frogs, *Pelophylax ridibundus* (*Rana ridibunda*), which hibernate underwater can tolerate the freezing of up to 55% of their body water although they do not produce an anti-freeze.

BELOW Spring peepers, *Pseudacris crucifer*, from North America can tolerate freezing conditions.

In conclusion, frogs are able to overcome conditions that we might consider to be unsuitable. They use a number of physiological adaptations that allow them to live in places that would appear, at first glance, to be too dry, too hot or too cold. In some cases they have turned these apparent drawbacks into advantages, and have moved into habitats, such as deserts, in which they face less competition.

4 Enemies and defence

WHEREVER FROGS OCCUR IN SIGNIFICANT NUMBERS, they are eaten by a wide variety of other animals. Their tendency to congregate together at certain times, especially during the breeding season, and their soft, relatively defenceless bodies, makes them vulnerable to predators, some of which specialize in frogs and eat little or nothing else, and others that eat them only on rare occasions, as the opportunity arises. Opportunists include such unlikely enemies as centipedes, freshwater crabs, hunting spiders, orb-weaving spiders and mygalomorph spiders (tarantulas) as well as larger animals. Frogs have even been found inside the traps of insectivorous plants such as the tropical pitcher plants, *Nepenthes* spp., and the Venus flytrap, *Dionaea muscipula*.

OPPOSITE Like many arboreal frog species, the Demerara Falls tree frog, *Hypsiboas cinerascens*, from Ecuador is well camouflaged when resting among leaves.

Types of predators

Frog-eating specialists include many aquatic and semi-aquatic snakes that hunt among the dense populations of frogs around the edges of ponds and lakes. Species such as the North American garter snakes and water snakes, *Thamnophis* and *Nerodia* spp., the European grass snake and related *Natrix* species, tend to eat mostly ranid frogs, whereas the African bush snakes, *Philothamnus* spp., eat mostly reed frogs, *Hyperolius* spp.. Basically, snakes eat whichever species are most abundant and adapt their hunting strategies accordingly. The American hognose snakes, *Heterodon* spp., and the African night adders, *Causus* spp., are toad-eaters and the hognose snakes, in particular, are experts at using their upturned snouts to dig toads out of the ground. Tropical snakes prey on a wider range of frogs, including tree frogs, frogs from the leaf-litter, and burrowing frogs, as well as aquatic and semi-aquatic species. Frogs that are sleeping are most vulnerable and nocturnal snakes tend to target diurnal frogs and vice versa. Leaving aside frog specialists, almost any snake will eat almost any frog, provided it can catch, overpower and swallow it. A number of snakes, like the American rat snakes,

BELOW European grass snakes, *Natrix natrix*, prey heavily on amphibians.

ABOVE Small frogs are often eaten by larger more powerful species such as the giant river frog, *Limnonectes leporinus*, from Southeast Asia.

Pantherophis spp., the cottonmouth, *Agkistrodon piscivorus,* and several elapids, vipers and pit vipers, may eat frogs as juveniles but change to different prey later in life.

Apart from snakes, in some places the most important predators might well be other frogs. Large species such as the horned frogs, *Ceratophrys* spp., in South America, the bullfrog, *Lithobates catesbeianus* (*Rana catesbeiana*) in North America, the African bullfrogs, *Pyxicephalus adspersus* and *P. edulis*, several of the wide-mouthed Australian *Litoria* (*Cyclorana*) spp., and barred frogs, *Mixophyes* spp., also from Australasia, and several large ranids from Southeast Asia, are all voracious predators that easily overpower smaller species. Some will also eat smaller members of their own kind and are therefore truly cannibalistic. The strange horned tree frogs, *Hemiphractus* spp., from South America, although relatively small themselves, have huge gapes and eat large numbers of frogs that are smaller than themselves.

Water birds such as herons, egrets, storks and ibises hunt frogs in shallow water where they are often abundant. In Asia and Madagascar, rice paddy fields and shallow reservoirs, or tanks, are good hunting grounds and few such places are without a population of wading birds. Other avian predators include members of the crow family, small hawks and some of the larger passerine birds, although it is doubtful if these species make great inroads into frog populations except perhaps at times when newly metamorphosed young are migrating away from ponds en masse. Nocturnal mammals, such as racoons, various monkeys, lemurs, mongooses, possums and hedgehogs are effective predators, while two genera of bats, the false vampire bats, *Megaderma* spp., from Asia, and the Central and South American fringe-lipped bat, *Trachops cirrhosus*, are frog specialists, using their accurate echolocation to home in on calling frogs even in total darkness. In urban and suburban areas, domestic and feral cats are significant predators of frogs. Other vertebrate predators include turtles, crocodilians and fish.

As larvae, frogs are equally vulnerable – perhaps more so. Juvenile water snakes that eat frogs when adult will prey on tadpoles in shallow water, and Central American cat-eyed snakes, *Leptodeira* spp., even eat the eggs of leaf frogs, *Agalychnis* and *Phyllomedusa* spp., that have been laid on leaves overhanging pools specifically to avoid aquatic predators. Fish, aquatic salamanders (newts) and their larvae eat frog tadpoles, sometimes wiping out the results of a whole season's reproduction. Frogs also eat eggs and tadpoles, including those of their own kind. Tadpoles of some species, such as the South American horned frogs, *Ceratophrys* spp., spadefoot toads, *Spea* spp., Cuban tree frogs, *Osteopilus septentrionalis*, and the poison dart frogs (Dendrobatidae) can be predatory on other tadpoles or on their own species, especially when they are overcrowded or during periods when other food is in short supply. Predatory aquatic invertebrates and their larvae are often present in small, temporary bodies of water that do not contain larger predators such as fish. Dragonfly larvae have an almost global distribution and are voracious predators, as are raft spiders, *Dolomedes* spp., although in terms of numbers their depredations are probably not critical.

LEFT A raft spider, *Dolomedes fimbriatus*, eating a common toad tadpole.

Defence

Collectively, frogs employ a wide range of strategies to avoid predation. Often, this is based on their particular habitat and life-style – what works for an arboreal species may not work, or even be possible, in the case of aquatic species for instance. Similarly, brightly coloured toxic species can safely be diurnal and indeed, their defence only works effectively if they can be seen easily and from a distance, giving predators enough time to identify and reject them.

Generally speaking though, the best way to avoid predators is not to be seen and frogs do this by hiding, or by looking like their surroundings. Being mostly nocturnal helps, of course, and during the daytime frogs hide under rocks, logs, leaf-litter or other debris. They also use leaf axils, crevices behind bark or a wide variety of other micro-habitats to prevent predators from finding them. In certain habitats, these situations also help them to avoid desiccation. Frogs that hide tend to seek out places where as much of their bodies as possible are in contact with a solid surface because, by jamming themselves into a small space, anything larger will not be able to follow them there (except snakes, of course, which are equally expert at getting into small spaces).

RIGHT A juvenile spotted reed frog, *Hyperolius puncticulatus*, hides in an arum flower.

Camouflage

Species that attempt to blend into their surroundings may simply have colours that match the substrates on which they rest. Thus ground-dwelling species are usually brown whereas tree and bush-dwelling ones are often green. These species are rarely of uniform colour but patterns of spots, blotches, bars and stripes help them blend into a background that is equally varied in hue. Leaf-litter frogs often have apparently random blotches in browns, creams and greys, imitating the various stages of leaf decay, and green tree frogs may have small rounded spots that mimic the holes and wounds made to living leaves by insects.

Markings consisting of blotches, stripes or chevrons on the frogs' bodies and limbs may not match the background exactly but they break up the outline of the frog and make it difficult for predators to find. Species with disruptive coloration of this kind are found among most families. They usually have elongated blotches on the sides of their heads, or dark lines that pass through the eye to form a mask-like marking that disguises the conspicuous outline of their eye. The markings are often continued onto the iris itself. Similarly, pale lines that start at the frog's snout, pass between its eyes and continue down its back, divide its shape into two halves which again, can make its outline difficult to recognize. Dorsal stripes are common to many species from several families, including many toads, Bufonidae, and many ranids. There is a recurring trend in many species for populations to be dimorphic, with some individuals, often about half, having a vertebral stripe whereas others do not. This disruptive coloration relies on the principal that predators have a mental image of the prey they are searching for and may overlook shapes that do not conform to this image. Some species are even more variable than this,

BELOW LEFT A ground-dwelling *Gephyromantis* spp., possibly *G. cornutus*. It is almost invisible in its natural habitat on the forest floor in Madagascar.

BELOW RIGHT The purpose of the vertebral stripe found on many frogs such as this Mascarene grass frog, *Ptychadena mascareniensis*, is to break up the outline of the frog. (Compare this with the photograph of the plain form of the same species on p.173.)

Polymorphism

Polymorphism is the occurrence within a population of two distinct forms, usually, but not necessarily, in colour or pattern. Different geographical races or subspecies are not polymorphic: the two forms have to occur side by side. Nor should it be confused with slight differences in colour and markings that are always present in any population of animals, nor with differences between males and females, which is sexual dimorphism. Finally, polymorphism does not include occasional mutations, such as albinos, unless they become fixed in a population and occur at a more-or-less constant proportion of the population.

Polymorphism can be advantageous to a population where predators build up a search image for a type of prey. Individuals that do not match the search image may be overlooked. If one type becomes more common than the other, more predators may build up search images for that type until its numbers are reduced and the other type becomes more common, and more heavily predated. Eventually, an equilibrium will be reached. All other things being equal, we would expect two or more types in a polymorphic population to occur in roughly equal numbers. There are many examples of polymorphic populations in frogs; a common variation is for populations to consist of individuals with or without pale lines running down the middle of their backs.

BELOW The Madagascar reed frog, *Heterixalus madagascariensis*, occurs in many colour forms. The bright yellow individual (below left) was living in the same tree as the pale blue one (below) along with several other variants.

with three or four colour forms occurring side by side. The argus reed frog, *Hyperolius argus*, is unusual in having males and females that look completely different from each other, with females having large cream spots and bars on a dark brown background whereas males are uniform pale green.

Camouflage is sometimes enhanced by the addition of skin flaps on the bodies and limbs of frogs. In its simplest form, this consists of small triangular processes or spines on the heels, and sometimes on the elbows as well. Other species, such as the frilled tree frog, *Rhacophorus appendiculatus*, have a scalloped fringe of skin along their flanks, chin and the outer edges of their limbs, also intended to break up the outline and confuse predators and, in some cases, these frills help to eliminate shadows by smoothing out the sharp-edged division between the raised edges of a frog and the surface on which it is resting.

Certain frogs take camouflage to extremes, often resulting in seemingly bizarre shapes and colours. Asian horned frogs, such as *Megophrys nasuta*, are not only coloured and shaped like dead leaves but also have leaf 'veins' on their backs and stem-like projections over their snouts and eyes. They even have small tubercles on their backs that look like the damage caused by leaf galls. In the Solomon Islands, another leaf frog, *Ceratobatrachus guentheri*, sometimes called the triangle frog, is similar, although it is not closely related to its Asian counterpart (see p.174). This species varies from reddish-brown to yellow in colour. An important aspect of both these species' behaviour is that they remain motionless in the presence of a predator

unless they sense they have been discovered, when they employ a secondary tactic, which depends on the species. The mossy frog, *Theloderma corticale*, which comes from Vietnam, has as its name suggests, the colour and texture of moss and is almost impossible to see when resting motionless on its chosen substrate. Even its eyes are

ABOVE The frilled tree frog, *Rhacophorus appendiculatus*, has flaps of skin that help to disguise its outline.

LEFT The Asian horned frog, *Megophrys nasuta*, is a master of disguise.

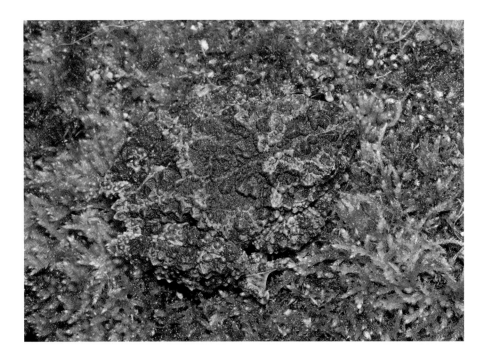

RIGHT The American grey tree
frog, *Hyla versicolor*, is difficult
to see when resting on bark.

mottled green. In Central America, Lancaster's tree frog, *Isthmohyla lancasteri* (*Hyla lancasteri*), is similarly coloured and textured. The North American grey tree frogs, *Hyla versicolor* and *H. chrysoscelis*, and the canyon tree frog, *H. arenicolor*, and a number of small frogs from Madagascar, including *Gephyromantis webbi*, and others, are mottled in greys and greens like lichen and habitually rest on the trunks of mature trees, while frogs with many light and dark longitudinal stripes are often found – or indeed *not* found – among grasses and reeds.

Escape

Another widely used line of defence is speed and agility. Frogs that live at the water's edge, for example, may not wait around to see if a predator has spotted them but will leap straight into the water at the slightest hint of danger and dive to the bottom. A walk along a frog-infested river bank, stream or pond-side will produce a succession of splashes, often followed a few minutes later by an equivalent number of pairs of eyes at the surface, as the frogs look to see if the coast is clear so that they can resume basking or feeding. Frogs of this type typically have streamlined bodies and long hind legs and they may also have pointed snouts. Among the most accomplished leapers are the African grass frogs, *Ptychadena* spp., but many frogs belonging to a number of different families are similar. The skittering frog (sometimes called the skipping frog) *Euphlyctis cyanophlyctis*, from Central and southern Asia, has an interesting variation on this behaviour. It usually rests at the water's edge and if disturbed it remains on the surface but pushes with its large hind feet so that it skims across the surface to the centre of the pond. As a last resort it will dive to the bottom. And in the far north of Australia, the rockhole frog, *Litoria meiriana*, also has the ability to skip across the water's surface, as does the rock skipper, *Staurois latopalmatus*, from Borneo.

Tree frogs of many kinds jump from branch to branch or may simply drop to the ground, while a few so-called flying frogs launch themselves from high branches and glide down to lower ones by spreading their toes and using the extensive webbing between

BELOW Stream-dwelling rock skippers, *Staurois latopalmatus*, can skitter across the water's surface to escape.

them as parachutes. These frogs have sticky toes that instantly grip any smooth surface they come into contact with, and the frog quickly hauls itself into a more comfortable position. Many tree and leaf frogs have flash markings on the thighs or other hidden surfaces, the purpose of which is to mislead predators. The markings, which are often yellow or red, with or without black spots and bars, are exposed for a brief period as the frog straightens its legs but are hidden the instant it comes to rest, leaving a confused predator looking for a splash of colour that is no longer visible. Examples of species with flash markings include several of the leaf frogs, *Phyllomedusa* and *Agalychnis* spp., and the red-legged kassina, *Kassina maculata*.

Feigning death (thanatosis)

If flight or concealment are not working, some species play dead, a natural phenomenon known as thanatosis. Hasselt's litter frog, *Leptobrachium hasseltii*, will lie on its back with its limbs pulled up under its body as though it has been dead for some time and has begun to shrivel up, and a similar position is adopted by some (perhaps all) of the bug-eyed frogs, *Theloderma* spp., from Asia and a number of tree frogs. Displays of this sort may last for just a few seconds before they right themselves and try to escape, or the frogs may remain motionless for a minute or more. We can only speculate on the purpose and effectiveness of thanatosis but have to assume that, in some cases at least it works, otherwise it would not have evolved. Some predators are stimulated by movement and ignore stationery items. Birds picking through leaf-litter for instance, probably rely on movement to detect prey so feigning death may be effective. Similarly, some arboreal species simply fall to the ground if they think they have been discovered and, by remaining motionless, the predator that disturbed them in the first place may be unable to relocate them.

RIGHT Mossy frogs and some other species sometimes feign death, a strategy known as thanatosis.

Toxic secretions

All frogs produce skin secretions from two types of glands in their skin: mucous glands that keep their skin moist, and granular glands. Granular glands appear to produce substances that inhibit fungal and bacterial infections and, in some species, also produce substances that are either distasteful or toxic to predators. These secretions contain a number of complex chemicals, some of which are unique to frogs. Their function is to make a predator drop the frog as soon as it picks it up in its mouth, but a number of species take this to extremes, producing highly toxic substances that can kill in minutes, if not seconds. They act in a variety of ways but may cause blood vessels to constrict, heart rates to increase dramatically, or block nerve impulses, causing muscles to contract leading to heart failure. They consist of a wide range of individual toxins that are often used in combination with each other and several toxins are named for the genus or species from which they were first isolated, such as bufotoxin, after the genus *Bufo*, pumiliotoxin, after the strawberry poison dart frog, *Oophaga pumilio* (*Dendrobates pumilio*), and histrionicotoxin after the harlequin poison dart frog, *O. histrionica* (*Dendrobates histrionicus*)

The distribution of the poison-producing glands varies. Toads belonging to the genus *Bufo* and related genera, for instance, are covered in warts of various sizes in places where there are heavy concentrations of these glands. Often, they have a large swelling behind each eye known as the parotid gland and this may be roughly circular or lozenge-shaped. Species such as the marine or cane toad, *Rhinella marina* (*Bufo marinus*),

BELOW The poison histrionicotoxin is named after the harlequin poison dart frog, *Oophaga histrionica*, of which this is just one form. Another is illustrated on p.30.

and the rococo toad, *R. schneideri* (*Bufo paracnemis*), are especially well-endowed, and have large parotid glands, elongated glands on the lower parts of their hind limbs (tibial glands) and many smaller patches of glands over their upper surfaces. In some species, such as the natterjack toad, *Epidalea calamita* (*Bufo calamita*), the glands are coloured differently from the rest of the toad's body. If any of these toads are roughly treated, they exude a white milky substance from the glands. If this gets into a predator's mouth it immediately drops the toad. If swallowed or absorbed into the tissues, it can make them violently ill, sometimes fatally. (A substance produced by the Colorado river toad, *Incilius alvarius* (*Bufo alvarius*), *O*-methyl-bufotenin, is a powerful hallucinogen and, for a while, toad-licking became a popular pastime with the experimentally inclined, but is not recommended.) A toad's reaction to danger is often to face the predator, lower its head and straighten its back legs, so that the large glands behind the eyes are immediately apparent and the first part of the body to be bitten or pecked. In doing so it may also make itself look bigger and more difficult to swallow.

RIGHT The glands of the Ranger's or raucous toad, *Amietophrynus rangeri*, (formerly *Bufo rangeri*), from South Africa are brightly coloured perhaps to draw attention to them.

RIGHT The toxins produced in the skin of the Colorado River toad, *Incilius alvarius*, have hallucinogenic properties.

LEFT *Trachycephalus* spp. from South America are known as milk frogs due to the large amounts of milky secretion they produce if handled roughly. This is the Amazonian milk frog, *Trachycephalus resinifictrix*.

Australian toadlets, genus *Uperoleia*, of which there are at least 26 species, also have greatly enlarged parotid glands and some have additional glands on theirs thighs and groin. South American *Pleurodema* species such as the Chile four-eyed frog, *P. thaul*, and the Cuyaba dwarf frog, *Eupemphix nattereri*, have concentrations of poison glands in raised areas on each side of the lower regions of their backs and these are known as inguinal glands. They are coloured black and white, contrasting with the rest of their dorsal coloration. Unlike most toads, they turn their backs on their enemy and raise their rumps to display these eyespots if they feel threatened. At first glance they can look like a pair of eyes. The South African Cape caco, or dainty frog, *Cacosternum capense*, also has prominent, raised inguinal glands, and another on its rump, which secrete toxins although this species does not seem to have a specific display. Tree frogs of the genus *Trachycephalus* secrete copious amounts of a substance that is poisonous and very sticky. The granular glands of these species are not concentrated into one or two areas but are distributed over the top of their head and its back.

Even more toxic are the poison dart frogs belonging to the family Dendrobatidae. About one third of this family – those belonging to the subfamily Dendrobatinae – are almost unbelievably colourful, with patterns that include large areas of yellow, red, green, orange or blue, sometimes in combination with each other but more commonly with contrasting black areas. They are diurnal and move boldly about on the forest floor or in the forest understorey, living in discrete colonies. These poison dart frogs produce some of the most effective toxins known. The golden poison dart frog, *Phyllobates terribilis*, is the most poisonous, producing enough homobatrachotoxin to kill several humans. This species is uniformly coloured bright orange and, at 35 mm (1 ⅓ in) in length, is one of the largest species in the family. Famously, a population of native Indians use the toxin from this species, and a couple of other species, to tip their blowgun darts.

ABOVE The granulated poison dart frog, *Oophaga granulifera*, is one of many dendrobatids whose bright colours warn of dangerously toxic skin sectretions.

In Madagascar the mantellas parallel the poison dart frogs. These species, which are not closely related to the dendrobatids but which belong to the family Mantellidae, are similarly coloured, and also toxic, though their secretions are apparently not as potent. Sixteen species are recognized at present although many populations are highly variable and future revisions may identify some of these as new species. A few of the *Atelopus* frogs from Central and South America are also brightly coloured and toxins have been isolated from their skins, the best known example being the Panamanian golden frog, *Atelopus zeteki*, one of several species in the genus that is critically endangered, and which produces a substance known as zetekotoxin.

RIGHT Mantellas, all of which are from Madagascar, parallel the American poison dart frogs in having toxic skin and bright coloration. This is Baron's mantella, *Mantella baroni*.

All the above species produce their toxins by sequestering alkaloids from prey animals, especially ants. The Australian corroboree toadlets, *Pseudophryne corroboree* and *P. pengilleyi*, from New South Wales, on the other hand, can produce their own toxins and, as such, they are unique among vertebrates as far as is known. The most important element is called pseudo-phrynamine and appears to be an effective protection against skin infections as well as deterring potential predators. Both these species are small, brightly coloured with black and yellow stripes, and both live in the mountains of New South Wales. Both are critically endangered, and the population of *P. corroboree* may number less than 100.

Other frogs produce strong poisons but do not advertise the fact quite as blatantly as the poison dart frogs, mantellas or corroboree frogs. The fire-bellied toads, *Bombina* spp. for instance, have bright yellow, orange or red areas of skin on their undersides and on the palms of their hands and feet but their backs are grey or green, for camouflage. When threatened, they arch their backs and raise their hands and feet to display the bright patches of skin. This behaviour is known as the Unkenreflex, Unken being the German name for fire-bellied toads. Other species with similar strategies include Nicholl's toadlet, *Metacrinia nichollsi*, and the sunset frog, *Spicospina flammocaerulea*, both in the Myobatrachidae, and the red-bellied toads, *Melanophryniscus*, from South America. All these are dark brown or black dorsally with large patches of yellow, orange or red on their undersides and on the palms of their hands.

In the predator–prey arms race though, frogs' poisonous skin secretions do not protect them completely from certain predators. Grass snakes, *Natrix natrix*, and hognose snakes, *Heterodon* spp., for instance, eat bufonid toads that other snakes avoid, while *Liophis epinephelus* eats poison dart frogs, including the highly toxic *Phyllobates terribilis*, although probably only the juveniles. Other predators, such as crows and ravens, have learned to eviscerate toads, eat their innards and leave the skin.

Mimicry

Mimicry has not been widely studied in frogs but there are a few presumed or documented instances. The harmless robber frog, *Pristimantis gaigei*, appears to be a Batesian mimic of the almost identical but highly toxic poison dart frog, *Phyllobates lugubris*, where they occur together in parts of Panama. A more complex situation involves the small imitator poison dart frog, *Ranitomeya imitator* (*Dendrobates imitator*), from the Andean foothills in Peru. This species occurs in at least three colour forms.

TOP The markings of the Oriental fire-bellied toad, *Bombina orientalis*, combine camouflage coloration on the dorsal surface with warning coloration on the underside.

ABOVE The southern corroboree frog, *Pseudophryne corroboree*, from New South Wales, Australia, lives in sphagnum bogs during the breeding season and lays its eggs in waterlogged moss.

Batesian and Müllerian mimicry

When a species evolves similar coloration and behaviour to another in order to benefit it, it is said to be a mimic and the species it mimics is the model. Mimicry is divided into Müllerian mimicry, named after F. Müller, and Batesian mimicry, after H.W. Bates. In Müllerian mimicry two or more distasteful or poisonous species benefit mutually by looking like each other because predators learn to avoid both after encountering either one of them. In Batesian mimicry, on the other hand, only one species is poisonous or distasteful and the other 'cheats'

ABOVE The red-headed poison dart frog, *Ranitomeya fantastica* (above left), and the imitator poison dart frog (above), are Mullerian mimics.

by looking similar to it despite being harmless. Batesian mimics should theoretically occur in smaller numbers than their models, otherwise predators may not learn to avoid them because they are more likely to encounter the harmless individuals than the distasteful ones, and this will be to the disadvantage of both species. There are examples of both types of mimicry in frogs.

Each form mimics another species of poison dart frog where their range overlaps: the variable poison dart frog, *Ranitomeya variabilis* (*Dendrobates variabilis*), the red-headed poison dart frog, *R. fantastica* (*D. fantasticus*), and the Amazonian poison dart frog, *R. ventrimaculata* (*D. ventrimaculata*). Each species benefits from looking like the other and they are, therefore, Müllerian mimics.

Smell

Even to the limited sensibility of the human nose, a number of frogs give off characteristic smells, which are probably associated with toxic secretions but which might also help to put predators off. The well-named skunk frog, *Aromobates nocturnus*, speaks for itself. First described in 1991 from Venezuela, it is related to the poison dart frogs and is critically endangered. The common European spadefoot toad, *Pelobates fuscus*, smells of garlic if it is handled and is also known as the garlic toad. Staying on a culinary theme, the Asian forest toad, *Phrynoidis aspera* (*Bufo asper*) has an odour reminiscent of peanuts.

Active defence (attack)

The image of frogs as more-or-less defenceless targets for a wide variety of predators holds true for the vast majority of species, which is why they go to such great lengths to avoid direct confrontation. A few species are prepared to fight back however, and these tend to be the larger, more belligerent frogs. Horned tree frogs, *Hemiphractus* spp., open their mouths widely to display the bright yellow interior and may snap at anything that comes within range. The helmeted water toad, *Calyptocephalella gayi*, has a similar display but also inflates its body and raises itself up on its legs to appear bigger. African bullfrogs, *Pyxicephalus* spp., simply lunge and bite, as do horned frogs, *Ceratophrys* spp., which can give a painful bite due to large tusks on their lower jaws. The closely-related Budgett's frogs, *Lepidobatrachus* spp., from South America, puff themselves up, stand stiff-legged and scream, with their mouths wide open. If this fails to have the desired effect they too will launch themselves at their tormentor. Screaming loudly is also common to a number of smaller frogs and is often enough to startle a predator into dropping them.

Defence in eggs and tadpoles

Frogs' eggs have no real defence except being laid in concealed places, or places where most predators cannot reach them, such as on leaves and branches hanging over pools. Some frogs guard their eggs, as discussed elsewhere (see p.105). Others invest their energies into producing such vast numbers that some are almost bound to survive despite a high degree of predation. Eggs of some species, such as *Atelopus* toads and some other bufonid toads are noxious and are usually avoided by predators. They obtain their toxins from their mother but as they feed and grow they dilute the chemicals and the protection may diminish or disappear altogether.

Tadpoles may rely on flight to escape from predators. Whereas many species are completely oblivious to disturbance and simply continue to swim slowly and in a seemingly aimless fashion, others dart to the bottom of the pool and may hide under stones or debris. Some are well camouflaged and difficult to see underwater, in contrast to the black tadpoles, produced by most of the familiar European and North American frogs and toads. Species that produce toxic secretions from their granular glands as adults also benefit from these secretions in the latter stages of their development as these glands begin to form, although even highly toxic species such as the poison dart frogs, do not fully develop their characteristic warning colorations until they metamorphose. Finally, the tadpoles of a few species form large aggregations or balls that appear to roll across the bottoms of the ponds in which they live. This may be a form of anti-predator behaviour, as it is in shoaling fish.

Summary

Frogs and toads use a variety of methods to avoid predation, by escaping detection, leaping or swimming away, intimidating their enemies or producing some of the most potent skin toxins in the natural world. All these strategies are successful because natural selection will have eliminated those whose defences were not adequate, but many frogs and toads rely on the sheer weight of numbers of offspring to ensure that, even with heavy predation, enough individuals remain to form a new generation.

5 Food and feeding

MOST TADPOLES ARE HERBIVORES, AT LEAST TO START WITH, grazing on algae, bacteria and higher plants, but adult frogs are carnivorous. They eat a range of prey, mostly invertebrates such as insects but also, in a few cases, vertebrates, including smaller frogs, small mammals and reptiles.

ABOVE Fry's frog, *Australochaperina fryi*, from Queensland, Australia is a small leaf-litter dwelling microhylid that eats large numbers of small prey.

Types of prey

Frogs are opportunistic and simply eat whatever they can catch, overpower and swallow. Their choice of diet depends on two main factors: the availability of different prey species and the size of the frog in question. Size is an important factor because they cannot dismember their prey – small species can only eat small prey. Prey availability depends on where they live and some frogs depend heavily on one or two types of prey whereas others eat a wider variety. In habitats with distinct seasons, the variety of available prey tends to be greatest during the wetter and warmer parts of the year and this is also reflected in frogs' diets. Very small frogs, especially microhylids that have small mouths and are sometimes known as narrow-mouthed frogs, eat a high proportion of leaf-litter mites, which can account for almost half the stomach contents of certain species. It is likely that mites probably figure highly in the diets of many small frogs, but this has not been studied to any great extent. Other invertebrates, such as springtails, or collembolans, which are equally small, occur in high densities in leaf-litter and are also eaten by small frogs, including juveniles, while ants and termites form a staple diet for a large number of species, especially burrowing frogs, such as microhylids, African rain frogs, *Breviceps* spp., Australian burrowing frogs such as the turtle frog, *Myobatrachus gouldii*, and many of the Australian toadlets, *Crinia* and *Pseudophryne* spp. The strange Mexican burrowing frog, *Rhinophrynus dorsalis*, eats nothing but termites and ant larvae. Poison dart frogs eat large numbers of ants and produce their toxic secretions by accumulating the chemicals found in them (and possibly other noxious invertebrates)

OPPOSITE South Amercian horned frogs, *Ceratophrys* spp., are predatory frogs. They have squat bodies and wide mouths enabling them to overpower and swallow large prey.

ABOVE Springtails (Collembola) and similar invertebrates form a large part of the diet of small frogs.

and the mantellas of Madagascar parallel this diet, as well as the bright coloration and toxic skin secretions that are associated with it. Ant-eating frogs tend to have small gapes and they are often active feeders, but some of them station themselves next to ant trails and pick off the individual insects as they pass by.

Larger frogs have many more options open to them. Insects often abound in the areas where frogs are numerous and these make up the bulk of the diet of the vast majority of species, not because of specialization but because they are the most likely prey to be available.

ABOVE Ants and termites figure heavily in the diets of some frog species such as Lea's frog, *Geocrinia leai*, from Western Australia.

BELOW South American horned frogs, *Ceratophrys* spp., have huge gapes and can swallow relatively large prey.

Crickets, grasshoppers, spiders and beetles figure largely in the diets of these more generalist frogs. European common toads, *Bufo bufo*, and marine toads, *Rhinella marina* (*Bufo marinus*) in Australia, have been known to sit at the entrance to bee hives snapping up bees as they come and go, although some individuals avoid stinging insects. Large frogs with wide gapes also eat small vertebrates: the American horned frogs, *Ceratophrys* spp., are particularly well-known for their voracious feeding habits, as are the African bullfrogs, *Pyxicephalus* spp., and all these species frequently eat smaller frogs as well as small reptiles and mammals that may come within range.

LEFT Kuhl's creek frog, *Limnonectes kuhlii*, eats freshwater crabs along with a variety of other food including other frogs.

The highly aquatic frogs belonging to the Pipidae feed underwater and the larger species take small fish. One member of the family, the African clawed frog, *Xenopus laevis*, is such an efficient predator that its presence in a lake can have a significant impact on local fisheries. Terrestrial frogs may also eat fish and a West African species, *Aubria subsigillata*, related to the African bullfrogs, apparently specializes in them. Other frogs with unusual diets include the snail- and slug-eating reed frogs, *Paracassina kounhiensis* and *P. obscura* (formerly *Tonierella kounhiensis* and *T. obscura*), from Ethiopia, which have modified skulls that allow them to open their mouth more widely and give them a more powerful bite so that they can more easily pluck molluscs from surfaces. They eat slugs and snails whole, including their shells if present, and the former species has elongated, curved teeth to help it grip slippery prey. Another African species, the Cameroon forest tree frog, *Leptopelis brevirostris*, is also reported to feed exclusively on snails but details of how it deals with them are lacking. The crab-eating frog, *Fejervarya cancrivora* (*Rana cancrivora*), eats crabs, as both its common and scientific names suggest. This species, which lives in Southeast Asia and has been introduced into India, has a high tolerance of salty water and hunts among mangrove forests and on estuarine mudflats at low tide. It does not eat crabs exclusively however, and in freshwater habitats where it also lives it eats a wide variety of other

BELOW The marine or cane toad, *Rhinella marina*, is one of the few species that will eat food that is not moving.

invertebrates. A small group of related swamp and river frogs from Borneo, *Limnonectes ibanorum*, *L. ingeri*, *L. leporinus* and probably *L. kuhli* also eat crabs, but of the freshwater variety, and it is only the larger adults, mostly males, that can manage them. These broad-headed species also eat a variety of other food, including smaller frogs.

There are very few frogs that will eat prey that is not moving although just a twitch of an antenna is usually enough to stimulate the feeding reflex. The marine or cane toad has learned to eat carrion and even kitchen scraps, rice, chicken bones and pet food.

Plant-eating frogs

As far as is known, only two species of frogs deliberately eat vegetable matter. Izecksohn's tree frog, *Xenohyla truncata* (*Hyla truncata*) from Brazil, eats the brightly coloured berries of a variety of plants and plays an important role in seed dispersal. The frog lives in the water-holding centre of a terrestrial bromeliad plant, *Neoregelia cruenta*, and the seeds germinate there after first passing through its gut. Seeds can account for two-thirds of its diet by volume, especially in the wet season when many plants are fruiting, but it also eats invertebrates. The Indian green pond frog, *Euphlyctis hexadactylus* (*Rana hexadactyla*) starts life as an herbivorous tadpole, as do most frogs, then switches to an insectivorous diet when it metamorphoses. As it grows into an adult, its diet switches back again, and it eats the leaves of aquatic plants and strands of algae, which together can make up nearly 80% of its stomach contents by volume.

BELOW Izecksohn's tree frog, *Xenohyla truncata*, is the only frog known to eat berries, although its diet also includes insects.

Hunting methods

Methods of hunting vary according to the type of prey and the size and shape of the frog. Generalist species, which eat a variety of insects and other invertebrates, catch much of their food during their nocturnal foraging, taking them as the opportunities arise. Many also eat prey that share the nooks and crannies where they spend the day – a common toad that lives under a large flat rock on my rockery seems rarely, if ever, to leave her retreat and probably finds enough to eat from the worms and insects that find their way into the cavity in which she spends her time. Agile species may leap long distances to catch their prey and tree frogs can catch flying insects in mid-air. Frogs that feed on small prey such as ants, mites, springtails, etc. need to eat almost constantly and appear to forage frenetically in order to find enough to eat. These species can only survive in places such as rainforests where small insects abound among the leaf-litter on the forest floor and their food intake is limited only by how quickly they can hop around and gather them up. Larger species, which eat larger prey, may only eat occasionally, and many use ambush tactics to catch their food, lying in wait among dead leaves, or just below the surface of the water, until a likely victim comes within range. Many species of this type are well-camouflaged. In order to maximize their hunting success, they have enormous mouths and powerful jaws so that they can take in a wide range of prey, and the horned frogs, *Ceratophrys* spp., are popularly named pac-man frogs (after the computer game Pac-Man in which the characters 'eat' dots.) Juvenile members of these species may use their yellow or pale-coloured toes to lure prey to within range, by twitching them among the dead leaves or raising them above their backs.

Frogs' tongues are central to the way in which they capture their prey and, although they are all basically similar from an anatomical viewpoint, they differ in the way in which they are used. Frogs belonging to some of the oldest families, the Leiopelmatidae, Bombinatoridae and some of the spadefoot toads, have a short circular tongue attached to the floor of their mouth. (*Discoglossus*, the generic name of the painted frogs, means disc-tongue.) Members of some of the more recently evolved families, such as the glass frogs, Centrolenidae, and many tree frogs, Hylidae, have a similar type of tongue. To catch their prey, species of this type lunge or jump forward and simply grab it in their mouth. They may use their tongue to manoeuvre the prey once it is in their mouth but their ability to project their tongue is very limited. Most of the more recently evolved families however, have long tongues that are attached at the front of their mouth only. They can flip their tongue out rapidly, use its sticky tip to capture their prey and pull it back into their mouth in one smooth operation while rocking their body forward slightly. The degree to which they can project their tongue varies between species but some toads, for instance, can extend it for a significant distance.

Microhylid frogs, and the shovel-nosed frogs from Africa, *Hemisus* spp., project their tongues in a different way. Hydrostatic muscles in the tongue can cause it to elongate suddenly, so that it shoots out of the frog's mouth like a rod, and can be accurately aimed at small prey. Studies on the African shovel-nosed frogs, *Hemisus* spp., and a large microhylid, the banded rubber frog, *Phrynomantis bifasciatus*, have shown that they can

Shovel-nosed frogs such as the marbled shovel-nosed frog, *Hemisus marmoratus,* from South Africa have unusual tongues that can be shot out rapidly using hydrostatic pressure.

project their tongue to either side as well as directly in front of their head so that, without turning, they can snap up prey within an arc of more than 180˚.

Members of the aquatic family Pipidae have no tongues. The small species feed on aquatic invertebrates but the larger ones eat larger prey, including fish. Their method of feeding is to wait until prey comes within range then to kick hard with their legs, so that their large webbed feet propel them forward at great speed. At the same time, they open their mouth and water rushes in, carrying the prey with it. Once they have it trapped in their jaws they may use their hands and fingers to stuff it into their mouth. Pipid frogs also have lateral line organs along their flanks that may help them detect the movement of potential prey and, in the *Pipa* species, there are small tactile sense organs at the tips of their fingers to help them to find prey when they are hunting in murky water or dense vegetation.

Tadpoles

Most tadpoles feed on plant material, including algae, bacteria and protozoans, which they graze from the surface of rocks, higher plants, submerged detritus or other surfaces. They have rows of teeth bordering their mouth, known as labial teeth, which they use to scrape away at edible material that then enters their mouth and is filtered out by food traps between their mouth and gills. Some tadpoles, such as those of the African clawed frogs and many others, skip the grazing stage and simply filter out minute food particles that are suspended in the water. Yet other species feed on small food particles floating on the water's surface, and some of these, such as the tadpoles of the horned frogs, *Megophrys* spp., have funnel-shaped mouths that enable them to harvest their food as efficiently as possible. The tadpoles of a South American microhylid, *Otophryne pyburni*, bury themselves in the sandy substrate of the streams in which they live and filter-feed on the small organic particles trapped between the grains. They have rows of long, narrow teeth that may act as sieves to prevent grains of sand entering their mouth. Tadpoles living in torrents and waterfalls have little opportunity to filter suspended material so they graze rocks. Some of these tadpoles even work their way up out of the water and into the splash zone where algae and bacteria grow more thickly. The tadpoles of *Amolops larutensis*, for example, climb several centimetres above the surface on mid-stream rocks. Similar behaviour occurs in African water frogs, *Petropedetes* spp., ghost frogs, *Heleophryne* spp., and others.

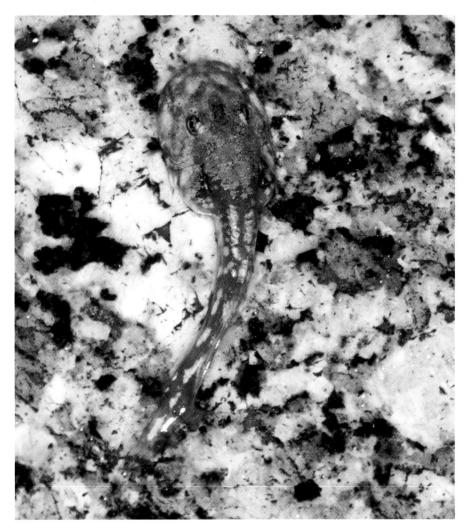

LEFT Tadpoles of the Larut Hills frog, *Amolops larutensis*, are adapted for clinging to rocks in fast-flowing streams and waterfalls.

ABOVE A tadpole of the Asian horned frog, *Megophrys nasuta*, showing the unusual upturned, funnel-shaped mouth with which it gathers food from the water's surface.

The tadpoles of most species eat animal material, including dead individuals of their own species, if the opportunity arises, and they are more likely to do so when food is in short supply. Although most of these tadpoles do not have any specialized mouthparts for eating meat, there are a small number of species that produce carnivorous tadpoles with wide heads and sharp teeth. Examples include those of the South American horned frogs, *Ceratophrys* spp. and the American spadefoot toads, *Spea* spp. The latter produce two types of tadpoles: the majority are typical grazers that feed on pond detritus but the other type has the potential to become carnivorous, with powerful jaws and serrated mouthparts. This carnivorous form only develops if there is a high density of small freshwater shrimps in the temporary pools in which they live but, having developed into carnivores, they may also take other tadpoles, including members of their own species, as well as the shrimps. African dwarf aquatic frogs, *Hymenochirus boettgeri*, produce predatory tadpoles of a different type, which have a tubular mouth that they use to suck up small invertebrates.

Frogs that breed in small bodies of water, especially inside the centres of bromeliad plants or tree-holes where food is limited, may produce tadpoles that eat the eggs of their own or other species. There are a number of these, including members of the Hylidae and the Rhacophoridae, but the best known are the poison dart frogs in the genus *Oophaga* (meaning egg-eater) such as the strawberry poison dart frog, *O. pumilio* (*Dendrobates pumilio*), the harlequin poison dart frog, *O. histrionica* (*D. histrionicus*) and a few close relatives. The females of these frogs carry their tadpoles to individual plants which they then visit every two or three days to lay an infertile food egg. Tadpoles of other poison dart frogs also eat the eggs of other frogs that might have bred in the same plant, or they may eat small invertebrates, or both. The Madagascan green-backed mantella, *Mantella laevigata*, appears to parallel the behaviour of the egg-eating poison dart frogs, although its breeding site is a tree-hole, usually a broken bamboo stem.

Finally, many species of tadpoles do not feed at all. These come from many different families and range from species that live in small bodies of water, as outlined above and develop in the normal way nourished only by their yolk sac, to species that remain inside the egg capsule until they are fully formed froglets, when they burst on the scene as perfect miniatures of their parents or, in a few cases, with just the remnants of their tails. Aquatic non-feeding types may have typical mouthparts identical to their relatives that feed in the normal way, or they may have lost their mouthparts, and this presumably reflects the length of time that has passed since they evolved from normal, feeding tadpoles.

BELOW The tadpoles of American spadefoot toads have the potential to become carnivorous under certain conditions.

6 Reproduction

REPRODUCTIVE BEHAVIOUR IN FROGS INVOLVES a number of different stages: mate attraction and choice, laying and fertilization of the eggs, subsequent development and metamorphosis. Each of these stages varies according to species and is also dependent on external factors resulting in a huge diversity of reproductive modes.

OPPOSITE The eggs of the golden mantella, *Mantella aurantiaca*, lack pigment because they are laid in a place where they are not exposed to bright sunlight, which could harm them.

Breeding seasons

In temperate parts of the world, such as northern and central Europe, North America and southern Australia, frogs reproduce in spring after a period of cool weather and possible hibernation. These seasonal breeders are ready to spawn as soon as conditions allow and they often migrate to their breeding sites within days of emerging from hibernation. Their eggs and tadpoles are thus given the maximum possible time to complete their development before the weather turns cold again in the autumn. Even so, tadpoles of a few northern species may overwinter and require a second, or in rare cases, a third summer in which to complete their development. In many seasonal breeders the whole population spawns within a few days, although timing may vary slightly from place to place depending on local conditions. Males arrive first at the breeding site and stay for the duration of the breeding season, leaving only when females stop arriving. Females make the journey to the site only to lay their eggs, which they usually do on the night of their arrival, and then leave, so on any given night there will be far more males than females present. This imbalance provides females with an opportunity to select which males to mate with.

A similar situation exists in the case of species that live in dry places and which can only breed when it rains. Activity is frenetic for one or two nights, by which time all the females will have laid eggs. These are known as explosive breeders. Depending on local conditions, there may be several bouts of breeding within a single rainy season, or the frogs may have to wait for a year or more before sufficient rain falls to form temporary pools again. In contrast to species from cool places, the tadpoles of these species develop rapidly so that they stand a good chance of metamorphosing and leaving the ponds before they dry up.

In wet tropical regions, conditions may be suitable for breeding throughout the whole year, or for the greater part of the year. Where there are one or more wet seasons, breeding activity is heightened at the onset of these, but may be extended for as long as the breeding sites have water, and many species breed opportunistically after any heavy rain storm. Tropical species that have direct development may breed continuously, and all stages may be present in the community at any given time as they are not dependent on rainfall provided the substrate remains moist.

Calling

BELOW An American toad, *Anaxyrus americanus* (formerly *Bufo americanus*), calling to potential mates. This species has a single, large vocal sac.

Breeding in most species is preceded by some kind of vocalization and although there are frogs that do not call they are very much in the minority. Males call by forcing air out of their lungs and across their vocal chords, into their mouth, where an inflatable part of the buccal cavity, the vocal sac, amplifies and transmits it to the outside world. The fire-bellied toads, *Bombina* spp., differ from other species because they produce their call by forcing air from their throat into the lungs. They do not benefit therefore, from the amplifying effect of a vocal sac, and the result is a quiet, repetitive 'ooop, ooop' call. The related painted frogs, *Discoglossus* spp., form a sound as air is passed in and out of the lungs. Male pipids do not have vocal chords and produce a series of clicking sounds underwater, sometimes likened to running a finger along the teeth of a comb.

Vocal sacs are located below the floor of the mouth and consist of a section of highly elastic skin. There are three basic types. The most common is a single balloon-like structure that can be almost as large as the frog's body in some species. In others, however, it is much smaller and, sometimes, barely noticeable. The loudness of the call is partially dependent on the size of the vocal sac, and species with very loud calls, such as many bufonid toads and some hylid tree frogs, have large vocal sacs. The high-pitched calls produced by many small tree frogs and reed frogs, for example, tend to travel further than low-pitched ones. In some species, the vocal sac is divided into two lobes, one on either side, with the central area inflating only a little, and this might help them to call when they are floating at the water's surface. There is also a third type, in which a pair of sacs are completely separate from each other and form at the angles of the jaw as in the European marsh frog, *Pelophylax ridibundus* (*Rana ridibunda*) and many other ranids.

Some species lack vocal sacs altogether and these include the fire-bellied toads mentioned above, and some of the Australian species belonging to the Myobatrachidae, such as the burrowing frogs, *Heleioporus* spp., in which the floor of the mouth performs the same function. Other species lacking vocal sacs include several bufonid toads from North America and a number of ranids from several regions. A number of frogs that breed near waterfalls and torrents do not call for instance, presumably because they would have difficulty in making themselves heard above the noise of rushing water, examples being the Australian waterfall frog, *Litoria nannotis*, and several hylid tree frogs from Central and South

ABOVE The European marsh frog, *Pelophylax ridibunda*, has paired vocal sacs and calls while floating at the water's surface.

America. Some torrent-dwelling frogs have other means of communicating, such as leg-waving, a form of behaviour observed in the Panamanian gold frog, *Atelopus zeteki*, the Borneo waterfall frog, *Staurois parvus*, and the Eungella torrent frog, *Taudactylus eungellensis*, from Queensland, Australia. Visual displays have been witnessed in other species of frogs and it seems likely that this form of communication is more widespread than was previously thought.

LEFT Male Australian waterfall tree frogs, *Litoria nannotis*, do not make mating calls.

Frogs call to advertise their presence to females and to other males. Females that are ready to breed are attracted to calling males, either in isolation or in choruses, while males may use their calls to space themselves out in order to avoid direct competition with each other for females. Some species call with a single repeated note whereas others include different elements, some of which are intended to attract females and others that warn off other males. The coqui frog, *Eleutherodactylus coqui*, from the Caribbean region, is one of the most studied species. The common name comes from its call, with the first part 'co' intended to repel other males while the 'qui' component attracts females. Some species, such as the rhacophorid tree frogs belonging to the genus *Theloderma*, have very complicated calls in which they seem to be communicating a variety of messages between each other, but the purpose of these vocalizations is not fully understood.

Calls also function as species-isolating mechanisms. They attract females to males of the same species and this avoids the possibility that the female will waste her eggs by allowing a male of a different species to fertilize them. Some groups or pairs of species are indistinguishable, at least to the human eye, but have different calls. These cryptic species were only discovered when scientists started to record and analyse the calls of frogs. The North American grey tree frogs, for example, were thought to be a single species, *Hyla*

BELOW The calls of the coqui frogs, *Eleutherodactylus* spp., have two functions, firstly to warn off other males and secondly to attract females.

versicolor, until the 1960s when scientists identified two distinct calls within the same populations. As a result, an additional species, Cope's grey tree frog, *Hyla chrysoscelis*, was described. Their ranges overlap in many parts of the eastern United States and the different calls serve to keep them separate. Similarly, the formerly widespread leopard frog, *Lithobates pipiens* (*Rana pipiens*) has been divided into several new species of which *L. pipiens* is but one, now known as the northern leopard frog. Some of the others are the southern leopard frog, *L. sphenocephalus*, and the plains leopard frog, *L. blairi*.

Males of some species call individually whereas others form choruses. Choruses tend to evolve in species that breed at the same time every year, in spring for instance, or at the beginning of a rainy season, and which concentrate in certain places for breeding such as a pond. By calling in choruses each is contributing to a sound that will carry further than that of a single frog and this probably helps to draw females to the breeding pond – choruses of some species can be heard over a kilometre away. The disadvantage of calling alongside many others of the same species is increased competition, and individual males may try to make their calls stand out from the others by being the first to start a chorus, and continuing after others have stopped. Only the strongest males can achieve this, which helps females to choose between potential mates; many males fail to find a mate.

Where several species use the same breeding site males of each species call from different positions to avoid competition. Males of one species may call in shallow water around the edge of a pond while others call from tussocks of grass at its fringes. Yet others may be calling from emergent vegetation such as reeds or from trees and bushes surrounding the pond. They may reduce competition still further by timing their calls to coincide with pauses in the choruses of other species, so that they do not interfere with each other. They may also call at different times, with some choruses beginning at dusk and others waiting until after dark.

Individual calling occurs when males are spread out in a habitat. This is usually associated with species that have scattered breeding sites. Males of tree-hole breeding

ABOVE LEFT The American grey tree frog, *Hyla versicolor*, shown here, is virtually indistinguishable from another species *H. chrysoscelis*, but their calls are different.

ABOVE RIGHT Names that describe a frog's call are often more useful than those that describe their appearance, no more so than for the motorbike frog, *Litoria moorei*, from Walpole, Western Australia, whose call sounds exactly like its name.

ABOVE Male Borneo tree-hole frogs, *Metaphrynella sundana*, call to attract mates to water-filled tree holes in which they lay their eggs. For reasons that are not clear this hole contains two individuals.

ABOVE RIGHT Male green bright-eyed frogs, *Boophis viridis*, call individually alongside forest streams in eastern Madagascar and do not form choruses.

RIGHT Male ornate nursery frogs, *Cophixalus ornatus*, from Queensland, call from hidden positions and stop calling at the slightest disturbance.

frogs for instance, of which there are species throughout the tropics, first find a suitable hole in which to breed and then call each evening to attract females. There may be several calling males in a relatively small area but they each have their own territory in the form of a water-filled tree hole, and females choose which to approach. In contrast to the situation with chorus-forming species, there is little or no direct competition between males. Similarly, species that lay their eggs in leaf-litter or moss do not form choruses even though there may be many males in a relatively small area, each with its own territory containing a calling site and one or more egg-laying sites.

One of the costs of calling is that it can attract predators as well as females. Some species of bats, as well as small mammals and larger frogs, may be attracted to calling males, especially those that form choruses. The males may adjust their calling rates as a response, by calling for a short time and then remaining silent for several minutes, to make it more difficult for potential predators to home in on them.

Mate selection and male–male competition

When a number of males of the same species call together, mate choice is often left to the female. As she approaches the calling site she orientates towards the male that is calling most loudly, or longest, or perhaps a combination of the two. Having made her choice she will approach him and initiate the mating process by allowing him to grasp her from behind in a position known as amplexus.

In some seasonally breeding frogs, the density of males is so high that females do not have an opportunity to select mates. Males tussle over all the newly arriving females in a scramble to obtain matings. Males that achieve amplexus are often unseated as other

LEFT Several male common toads, *Bufo bufo*, attempting to achieve amplexus with a single female, hidden in the centre of the melee.

ABOVE An amplectant pair of sunset frogs, *Spicospina flammocaerulea*, which is a rare species from Western Australia, is accompanied by a satellite male hoping to steal a mating.

males force their way between them and the female, and in some cases large mating balls completely overwhelm females, who drown without having laid their eggs. Scramble competition replaces female mate choice with a free for all in which the strongest males are the most successful. The distinction between female choice and scramble completion is not clear cut, however. Females will attempt to find the males with the most attractive calls but, when densities build up, are often unable to do so because other males intercept them before they reach the male of their choice. At some stage the system switches from female choice to scramble competition.

Indeed, some males intercept females while they are still en route to the breeding site and take up their amplectant position before they arrive, often to be ousted by larger males once the female enters the water. So-called satellite males position themselves near a calling male in the hope of intercepting a female before she reaches her chosen male. These may be smaller males that have little hope of competing with the louder calls of larger males, or they may be males that have called previously but switch tactics temporarily to conserve energy.

Amplexus

Fertilization in all but about 20–30 species of frogs is external – the eggs are fertilized by the male as the female releases them. This occurs during amplexus, the name given to the position frogs take up during the laying and fertilization of the eggs. Usually, the male sits quite high on the back of the female during amplexus and his forearms reach around and grasp her in the region of her armpits. This is known as axillary amplexus. In other cases, the male sits further back and grasps her around her waist, just in front of her hind limbs. This is known as inguinal amplexus. Males often develop roughened swellings on their hands or digits, and sometimes on their chests, during the breeding season, which help them to maintain their grip. The older families of frogs (Pipidae, Bombinatoridae, etc.) tend to perform inguinal amplexus, whereas the most recently evolved ones (Bufonidae, Ranidae, Hylidae, etc.) use axillary amplexus. The function of amplexus is to position the cloacae of the male and female close to each other so that the male can release the sperm to fertilize the eggs as the female is laying them. The male may arch his back at the moment of egg-laying to bring their cloacae together and, at the same time, use his hind legs to form a triangular area that holds the eggs for a few seconds so that the sperm has a better chance of reaching them. Without this precaution, sperm could be swept away in the water before it has reached all of the eggs, or the sperm of other males could fertilize some of them. Amplexus usually lasts for a few hours at most, but can be

LEFT Yellow-bellied puddle frogs, *Occidozyga laevis*, in axillary amplexus, Borneo.

ABOVE Asian horned frogs, *Megophrys nasuta*, in inguinal amplexus. Males of this species are significantly smaller than the females.

much longer in some species, notably some *Atelopus* species, which sometimes remain in amplexus for several months before egg-laying takes place. At the other extreme, amplexus in painted frogs, *Discoglossus* spp., lasts for only a few seconds and even less in some dendrobatids and mantellids.

There are species in which amplexus varies from either of the two positions described above. In some members of the Dendrobatidae, all of which lay their eggs on the land, the male turns his hands outwards and presses the back of them against the female's throat (cephalic amplexus), while in other members in the same family, there is hardly any contact at all as males and females sit tail to tail during egg-laying and fertilization. In some arboreal mantellid species, such as members of the genus *Guibemantis*, the male straddles the female's head while they are clinging to a vertical leaf or stem and fertilizes the eggs by releasing sperm which trickles down the leaf until it reaches the newly laid egg mass. Male rain frogs, *Breviceps* spp., are so rotund that their arms will not reach around the female's body and, instead, he secretes a substance from the skin on his underside that glues him to the female for up to three weeks; other frogs of similar shape also glue themselves together.

ABOVE Male tailed frogs achieve internal fertilization by using their tail, which is actually an extension of their cloaca.

A small number of frogs have internal fertilization. They are the tailed frogs, *Ascaphus montanus* and *A. truei*, from North America, in which males have an extension to their cloaca (the tail) with which to introduce sperm into the female's cloaca, at least two species in the genus *Eleutherodactylus* from Puerto Rico (one of which appears to be extinct) and in a number of small bufonid toads from Africa belonging to the genera *Altiphrynoides*, *Mertensophryne* and *Nectophrynoides*, none of which have common names. A number of these species give birth to live young, a topic that is discussed in Chapter 7.

The spawn

Female frogs produce spawn, a term that is used to describe the eggs (or ova) and the jelly mass that surrounds them. The number of eggs in a clutch varies tremendously, from less than ten in small frogs that display some kind of parental care such as poison dart frogs and some marsupial frogs, to well over 1,000 in larger species with little or no parental care. The marine or cane toad can lay up to 20,000 eggs in a single clutch and the American bullfrog about 10,000. Some species lay all their eggs in one spawning event whereas others lay smaller clutches at regular intervals.

When they are first laid, frogs' eggs are spherical structures that have an upper, animal pole that eventually divides and differentiates to form the embryo, and a vegetal pole that provides yolk for the developing embryo. The animal pole is often covered by a layer of melanin so that the eggs are black when seen from above. This is thought to protect the developing embryo from the harmful effects of ultra-violet rays and may also help to raise their temperature and so increase their rate of development. Eggs that are not exposed to sunlight, because they are laid under leaves, logs, moss etc., are usually white. Even among closely related frogs, such as the tree frogs belonging to the genus *Phyllomedusa*, those that wrap their eggs in a leaf have white eggs whereas those in which they are exposed are pigmented. Eggs of a few species of frogs that spawn on leaves, such as some of the glass frogs, Centrolenidae, and the New Guinea tree frog, *Litoria iris*, are greenish in colour, perhaps to camouflage them.

The eggs of many species are laid singly, scattered over the bottom of a pond or loosely attached to aquatic plants or detritus. The African clawed frogs, *Xenopus* spp., have this type of spawn. The related dwarf aquatic frogs, *Hymenochirus* spp., however, lay single eggs that float to the surface. Many frogs lay their spawn in clumps and again, they may be attached to aquatic vegetation or float just below the surface, as in the European common frog, *Rana temporaria*, where many clumps are often combined into a large raft of spawn, which is thought to absorb and conserve heat more efficiently and so speed up development. Frogs that breed in shallow temporary pools, such as many microhylids, deposit them in a single floating layer, probably because this layer is richer in oxygen in an otherwise stagnant pool. Bufonid toads lay their eggs in long single or double-stranded strings, wrapping them around water plants as they move about below the surface, while frogs that breed in fast-flowing streams and cascades, such as several of the *Hylarana* spp. from Southeast Asia, attach them to rocks, often on their undersides. Sticky outer capsules are characteristic of these species, to prevent them from being washed downstream.

·Foam nests

A different method of depositing spawn involves the construction of a foam nest, in which the albumen, which normally surrounds individual eggs, is whipped, or whisked, up as the eggs are laid so that instead of each egg having its own jelly-capsule, the jelly is combined into one foamy mass and the eggs are scattered throughout. Foam-nesters occur in several families and in various parts of the world. Some families consist exclusively of foam-nesters whereas others contain species that lay aquatic eggs and others that make foam nests. There is more than one type of foam nest and they differ in the way they are formed and where they are positioned.

The most common type of foam nest is made by males. The male whips up the albumen into foam with his hind legs as it is extruded by the female during egg-laying. In some species both the females and the males do this. Rhacophorid frogs from Southeast Asia, including members of the genera *Polypedates* and *Rhacophorus*, are foam-nesters, building their nests above the water on overhanging vegetation, a muddy bank or in a shallow puddle. The East African grey tree frog, *Chiromantis xerampelina*, also a rhacophorid, attaches its foam nest to the branch of a tree hanging over a waterhole. By the following morning the foam will have formed a crust, similar to that of a meringue, which it somewhat resembles, and this prevents the eggs and developing tadpoles from drying up in the intense heat. Members of the Leptodactylidae in Central and South America make foam nests similar to those of the rhacophorids. *Physalaemus pustulosus*, for example, makes a walnut-sized nest in shallow water. One

BELOW African grey tree frogs, *Chiromantis xerampelina*, build their foam nests over waterholes. Sometimes they form large mating aggregations, where many males contribute sperm and help to whip up the foam from one or more females.

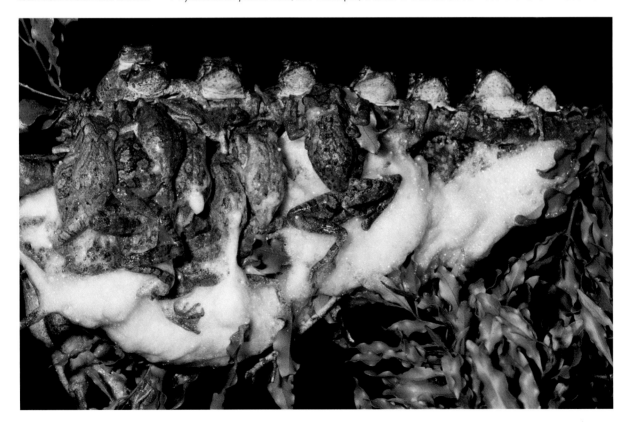

genus of microhylid frogs, *Stumpffia* spp. from Madagascar, makes a foam nest among leaf-litter on the forest floor.

Australian foam-nesters, which number over 20 species in the Limnodynastidae, make floating nests on the surface of the water, among inundated sphagnum moss or in flooded burrows. Here, the foam is made by the female that paddles the surface of the water with her front feet, which develop flanges along the digits during the breeding season. As she lays her eggs she pushes the bubbles under her body and between her hind legs to mix with the jelly mass and eggs. Foam nests of this kind are less cohesive and can only stay together because they float on still water.

Development

The transition from seemingly inanimate eggs to adult frogs with a full set of working parts is captivating to small children and scientists alike. Amphibians are the only vertebrates that go through an aquatic, water-breathing larval phase and metamorphose into terrestrial air-breathing adults: the word amphibian comes from two Greek words, 'amphi' and 'bios' that together mean 'both-lives'.

ABOVE A foam nest of the four-lined tree frog, *Polypedates leucomystax*, formed at the edge of a flooded wheel-rut.

Frogs' eggs begin to divide as soon as they are fertilized and go through several stages before hatching. The rate at which they develop is dependent on temperature, with those from cooler climates taking longer than tropical ones, although other factors such as the depth of water and the position of the egg-laying site also contribute to this. The sequence of events, as it applies to typical aquatic breeding species, is as follows, although there are some variations.

The round egg changes to a comma shape and shortly after this gills and mouthparts appear. The tadpole usually breaks free from its jelly-capsule at this stage and clings to the jelly mass or some other surface. The external gills become covered with a flap of skin called the operculum, which connects with the outside world through an opening called the spiracle, which is usually positioned on the left of the tadpole's body. The tadpole gulps water through its mouth, the water then passes across the gills, where gaseous exchange takes place, before leaving through the spiracle. Small particles of food in the water are filtered out by sieve-like structures between their mouth and gills. The hind limbs appear, followed by the fore-limbs, the left one emerging through the spiracle. The mouth changes shape until it is more frog-like and the tail begins to shrink as it is absorbed into the frog's body. The young frog may leave the water at this stage and the tail is completely absorbed a day or two later.

Tadpoles (the word comes from old English tadde meaning toad, and polle meaning head – so toad-head, because they look like large heads with a tail) are less variable than adult frogs. Most are black, dark brown or dark grey in colour, although some are speckled

Life-cycle of a frog

Nine stages in the life-cycle of the spotted reed frog, *Hyperolius puncticulatus*, from Tanzania, are illustrated below. The Gosner stages given in brackets refer to a standard method of describing the various stages of embryo and larval development, published by K. L. Gosner in 1960.

The top row of images, from left to right, show the eggs developing on a leaf overhanging water at day three (Gosner 18–19), then at day five (Gosner 20–21) and finally at day six (Gosner 22).

The middle row of images, again left to right, show the eggs developing at day eight just prior to hatching (Gosner 24), a single tadpole with hind limb buds (Gosner 34) and a tadpole with hind limbs and in which the coiled intestine is clearly visible (Gosner 37–38).

The last row of images show the final development stages of the tadpole with front and hind limbs (Gosner 42), a newly metamorphosed froglet (Gosner 46) and finally an adult frog.

After hatching, there is some variation in the rate of development but the average time taken from egg-laying to metamorphosis for this species is about 12 weeks. Frogs and toads from temperate regions develop more slowly however, whereas species that breed in temporary pools have evolved very short development times to avoid desiccation.

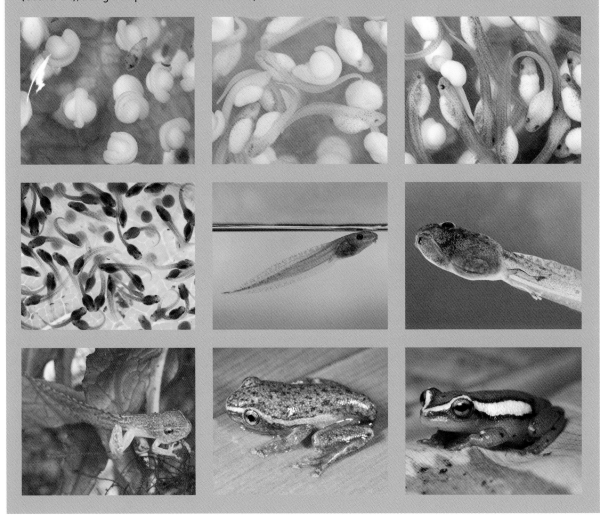

with dark and light markings and a few have patterns that help to camouflage them. Two basic types can be recognized: the pond-type, which has a plump body and high dorsal and ventral fins, and the stream type, which is more slender, has a muscular tail and lower fins. There are many variations as well as specialized tadpoles that have unusual habitats or life-styles. Most tadpoles are herbivorous and have scraping mouthparts, whereas a few are predatory, as described in Chapter 5.

Species from fast-flowing streams often have their mouths on the underside of their head, surrounded by a sucker-like structure that allows the tadpole to attach itself to rocks. It moves by creeping over the surface, rarely letting go completely, and by this means it can even climb out of the water onto the wet surfaces of emergent rocks and waterfalls. Tadpoles of this type are found in many families, including *Atelopus* spp. (Bufonidae), *Amolops* spp. (Ranidae), the ghost frogs (Heleophrynidae) and the tailed frogs, *Ascaphidae* (Leiopelmatidae). Other stream-dwellers, such as some of the *Colostethus* species, have tadpoles with funnel-shaped mouths that they can use to adhere to the underside of larger stones. Stream-dwellers may also live within the gravelly stream-bed and their elongated shapes are adaptations for moving through the small interspaces between stones.

Other elongated tadpoles are adapted to living in crevices. Some, such as certain hylid tree frogs and poison dart frogs, live in small bodies of water, such as those that collect in bromeliad plants and can wriggle across wet leaves if necessary, while the pandanus frogs from Madagascar, *Guibemantis* spp., which belong to the Mantellidae, develop in the leaf-axils of pandanus plants in exactly the same way and can flip themselves from one leaf axil to another.

Non-feeding tadpoles

Some tadpoles live in small bodies of water that do not contain enough food to sustain them. These might be hollows in the ground, tree-holes or the urn-shaped centres of bromeliads and other plants. Two alternatives to conventional feeding have evolved. Either the tadpoles eat the eggs of other frogs, or infertile eggs laid specially for them by their mother, as in the poison dart frogs of the genus *Oophaga* and several tree frogs, or they need to contain enough yolk to see them through their complete development until metamorphosis. Non-feeding tadpoles are found in many families and they show a complete spectrum of types, from species with normal mouthparts that feed if they have the opportunity but grow and develop without feeding if necessary, to species in which all traces of the feeding mouthparts have been lost. Tadpoles that do not feed are an evolutionary step towards terrestrial direct development and some of the more unusual forms of parental care in which one of the parents cares for, and sometimes carries eggs, tadpoles or both. These themes are explored in more detail in the following chapter.

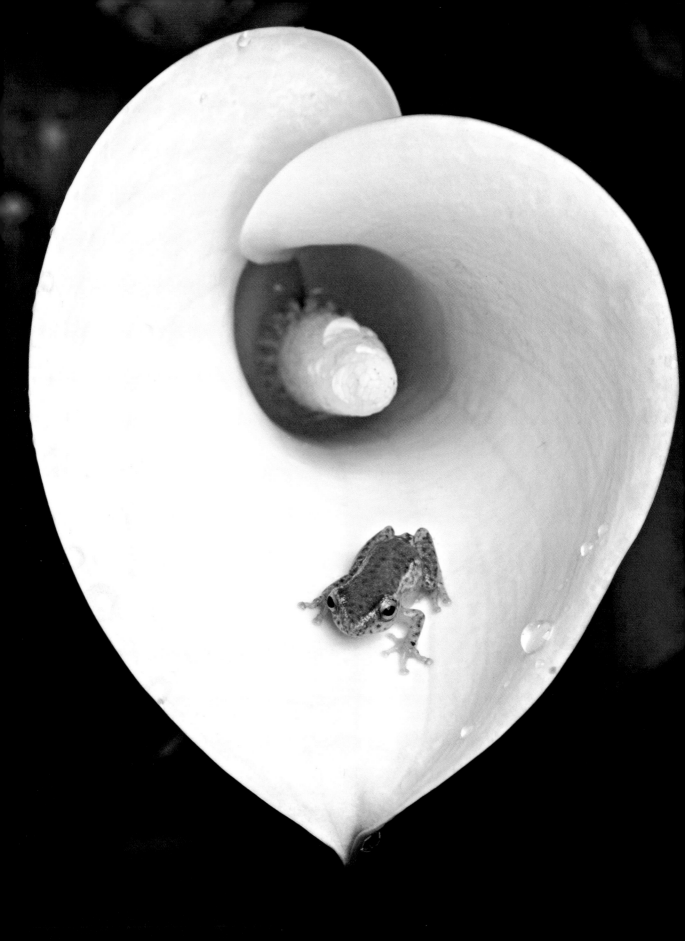

7 Life-cycles

THE PREVIOUS CHAPTER DESCRIBED THE TYPICAL PATTERN of egg-laying and subsequent development of tadpoles in the frogs that most people are familiar with. With this as a starting point, there are endless possibilities for variation in almost every stage of a frog's development, from the places in which they lay their eggs to the methods by which their tadpoles feed and develop. There are two important trends. The first is towards terrestrial development, which reduces the need for water to a minimum, so that frogs can breed in places where standing water is scarce or unreliable. The second is towards parental care, which gives eggs and offspring a greater chance of survival. These two factors often go hand in hand because clutches of eggs laid on the land are more easily watched over by the parents than those deposited in water. The production of non-feeding tadpoles, sustained only by their yolk, is another essential element in the evolution of parental care.

OPPOSITE A juvenile spotted reed frog, *Hyperolius puncticulatus*, sits in an arum flower, Tanzania.

With aquatic eggs and tadpoles

BELOW Desert toads such as the Sonoran toad, *Anaxyrus debilis*, which lives in the arid regions of North America, breed in temporary pools immediately following heavy rains.

It is generally assumed that the primitive method of breeding in frogs is fully aquatic. Females mate and lay their eggs in the water and the tadpoles continue their development there, until they metamorphose into young frogs. About half of all known frogs breed in medium to large bodies of permanent still water – ponds, swamps, marshes and lakes. Their eggs may be laid singly, in clumps or strings, they may be attached to aquatic vegetation or float at the surface, or they may be suspended in floating foam nests. All these are described in the previous chapter. The number of eggs varies with the species, with the larger ones tending to lay more eggs. In desert regions, species such as the spadefoot toads, *Scaphiopus* and *Spea* spp., the Mexican burrowing toad, *Rhinophrynus dorsalis*, the South African bullfrog, *Pyxicephalus adspersus*, and many others, breed in flooded areas immediately following heavy rain. The water surface may be extensive initially but soon evaporates leaving tadpoles in small, rapidly shrinking pools from which they must metamorphose quickly, in as little as 14 days in American spadefoot toads, *Scaphiopus* spp., and 18 days in the African bullfrogs, *Pyxicephalus adspersus*. Some species, such as the American spadefoot toads, *Spea* spp., produce tadpoles that can become predatory and cannibalistic if food becomes scarce, described in Chapter 5.

Frogs that breed in fast-flowing streams face different problems. They have to ensure that their eggs are not swept away and many species avoid this by using backwaters that collect at stream edges, or splash pools that form near waterfalls. Others, however, such as some of the torrent frogs from Asia, *Hylarana* spp., and from Australia, *Taudactylus* spp., attach their eggs to rocks on the streambed. Tadpoles of many stream-breeding frogs are equipped with suctorial discs on their undersides that allow them to cling to rocks while they feed. The tailed frogs, *Ascaphus* spp., the ghost frogs, *Heleophryne* spp., and the torrent frogs from Borneo, *Meristogenys* spp., are examples of the many frogs that have adaptations of this kind.

The laying of aquatic eggs leaves little opportunity for parental care and most species rely on large numbers to offset predation. There are some exceptions, though. The South American nest-building frogs belonging to the genus *Leptodactylus* make foam nests and, in some species at least, the female stays nearby. When the eggs hatch the tadpoles form a large shoal which the female herds together, and she may lead them to another

ABOVE The northern torrent frog, *Meristogenys orphnocnemis*, from Borneo lives and breeds in fast-flowing rainforest streams and is thought to attach its eggs to rocks on the streambed.

RIGHT The flattened tadpole of the Cape ghost frog, *Heleophryne purcelli*, has a suctorial disc on its underside so that it can cling to rock surfaces.

part of the pond, apparently communicating by means of pumping actions that send vibrations through the water. Females may defend tadpoles against predators. In the African bullfrog, *Pyxicephalus adspersus*, it is the male that stays near its tadpoles and he will defend them vigorously against predators, even leaping at humans and lions. If the pool in which its tadpoles are living begins to dry up, this species goes a stage further by digging a channel through the mud to connect the tadpoles' shrinking pool to a larger one containing more water.

Small water-depressions in the ground, such as hoof prints, the empty husks of forest fruits and nuts, seedpods, snail shells and even discarded tin cans, are used as breeding sites by a variety of frogs. Such small pools are often without predators such as fish (although they may contain predatory insect larvae) but they can be too small to support large numbers of tadpoles so they tend to be used by smaller frogs, which lay fewer eggs. Although there is a risk that they will dry out, frequent rain showers usually keep them topped up. The tadpoles of these species may feed on dead leaves and other debris or they may be non-feeding tadpoles nourished only by their yolk.

ABOVE **Male African bullfrogs,** *Pyxicephalus adspersus*, stay close to their brood of eggs and tadpoles and, if necessary, may force a channel between their drying pool and one that still contains water.

They include the Chilean ground frogs, *Eupsophus* spp., and the New Zealand frogs, *Leiopelma* spp., in which tadpoles remain in their small pools of water and develop without feeding. In the case of two of the New Zealand species, the male remains with the developing tadpoles and they may climb up onto his back, the purpose of which is not known. The tadpoles of Tschudi's froglet, *Crinia georgiana*, from Western Australia, which breeds in shallow water in bogs, have mouthparts and will eat if food is available but can grow and develop using only their yolk reserves if necessary.

LEFT **Tadpoles of the quacking or Tschudi's froglet,** *Crinia georgiana*, from Western Australia feed if they are able but will also develop without feeding if necessary.

Males of five species of Central and South American hylid frogs belonging to the genus *Hypsiboas* (formerly *Hyla*), known as gladiator frogs, build small shallow, basin-like pools at the edge of a forest stream or seep by scooping out mud to form a circular dam. The male calls from the edge of the pool to attract females and he may also have to fight off other males that try to take over his site. The eggs are laid inside the pool where they float on the surface and the male continues to guard the pool until the eggs have hatched, by which time the mud dam will have collapsed and the tadpoles can escape into the stream. In the stream-breeding giant river frog, *Limnonectes leporinus*, from Borneo and a related species, *L. blythii*, from Vietnam, the males and females work together to scrape a hollow in the gravel at the edge of forest streams, mixing their eggs with the sand or gravel that they dislodge.

Small pools that form in vegetation or in tree-holes are widely used as breeding sites in forest habitats. In Latin America many species make use of bromeliads, which can hold a substantial volume of water in their centres. In Jamaica, for example, all the hylid frogs use bromeliads, as do two species of bromeliad tree frogs, *Bromeliohyla bromeliacia* and *B. dendroscarta*, from Central America. In the Old World, where there are no bromeliads, frogs use water that collects in the leaf axils of forest plants such as pandanus. Tree holes and broken bamboo stems containing stagnant water are widely used in many parts of the tropics: by tree-hole frogs, *Metaphrynella pollicaris* and *M. sundana*, in Southeast Asia; a number of *Platypelis* species from Madagascar; and the Amazonian milk frog, *Trachycephalus resinifictrix*, from South America, for instance. In some cases the tadpoles do not feed but subsist on their yolk whereas in others the female provides food in the form of infertile eggs (see p.100). Holes in rotten logs are widely used by terrestrial species, such as the African bufonid toads, *Mertensophryne micranotis* and *M. anotis*, which also breed in water that collects between the buttress roots of large forest trees. The sticky frog, *Kalophrynus pleurostigma*, from Malaysia, normally uses hollows in rotting tree trunks but their tadpoles have also been found in the pitchers of pitcher plants, *Nepenthes*. Its tadpoles do not feed and lack mouthparts.

Terrestrial eggs and aquatic tadpoles

Frogs from many different families lay their eggs on land, implying that this method of reproduction has evolved independently in many families. Subsequent development may also be on land but, in many species, the tadpoles complete their development in water. There are many variations. In its simplest form this consists of eggs that are laid on ground that is later flooded. Some Australian *Geocrinia* spp. such as the roseate frog, *G. rosea*, and the toadlets, *Pseudophryne* spp, breed in this way. The latter are especially interesting because they lay their eggs in depressions or short burrows in the autumn and hatching is delayed until winter rains flood the site. Mantellas, *Mantella* spp., from Madagascar, lay their eggs in depressions in marshes and wet forests and subsequent rains flush the tadpoles into larger bodies of water, although some species in the genus use tree-holes. The African bush frogs, *Leptopelis* spp., lay their eggs in small depressions

near streams or ponds and, when they hatch, the elongated tadpoles wriggle from the nest into the water. The South American wood frogs, *Batrachyla* spp., from Chile and Argentina, appear to have a similar system. Shovel-nosed frogs, *Hemisus* spp., breed in underground chambers. When the eggs hatch, the female, which stays with the eggs during their development, tunnels her way through the mud and out into open water, with the tadpoles following her.

ABOVE The unusual grey wood frog, *Batrachyla leptopus*, from Central Chile lays its eggs on the ground and the tadpoles make their way to water when they hatch.

Laying eggs over water

In addition to the foam-nest builders, other frogs lay their eggs on rocks, roots or vegetation overhanging water, into which the tadpoles drop when they hatch. The glass frogs, family Centrolenidae, from Central and South America, lay their eggs on leaves overhanging forest streams. In some species, the male parent sits on the leaf with the eggs, sometimes leaving them during the day and returning at night, or sometimes

RIGHT Two male reticulated glass frogs, *Hyalinobatrachium valerioi*, guarding their clutches on a leaf overhanging a forest stream.

RIGHT A clutch of frog eggs, probably *Mantidactylus* spp., developing out of the water, Andasibe, Madagascar.

ABOVE African leaf-folding frogs such as Fornasini's spiny reed frog, *Afrixalus fornasini*, from East Africa lay their eggs above the water and fold a leaf around them.

staying there day and night, depending on the species. Some species have markings of round pale spots on their otherwise green bodies, which may be intended to mimic the egg-mass. The main predator of the eggs is a type of parasitic wasp that lays its eggs in the spawn, and it has been shown that those species in which the males stay with the eggs during the day tend to lose fewer eggs to the parasite than those species where males do not.

In Africa, several species of reed frogs, *Hyperolius* and related genera, also lay their eggs on leaves, stems and vines above the water. In *Hyperolius spinigularis* the female visits the eggs every night to squirt water from her bladder over them, while the females of *Alexteroon obstetricans*, which belongs to the same family, remain with their clutch and help the tadpoles to break free of the jelly when they are ready to hatch by kicking the jelly mass. Other frogs fold a leaf around the clutch, creating a tube in which the eggs are protected from dehydration and predation. The African leaf-folding frogs, *Afrixalus* spp., use this technique as do the large and showy South American leaf frogs, *Agalychnis*, *Pachymedusa* and *Phyllomedusa* spp. In *Phyllomedusa trinitatus* males call from a bush overhanging a pond. When a female approaches, the pair go into amplexus and move down to the tip of one of the outer leaves. As the female begins to lay she moves slowly up the leaf, gripping its outer edges with her back feet and pulling them over to form a tube around the clutch. Before she begins laying fertilized eggs she deposits a mass of jelly at the bottom of the tube and another empty mass at the top, so that both ends are plugged. The eggs hatch about six days later and their wriggling, coupled with the increasing liquidity of the jelly mass, allows them to drop from the leaf and into the water.

In a combination of adhesive eggs and tree-hole breeding, there are frogs that attach their eggs to the inside walls of tree-holes or other water-filled cavities. Several species of the warty tree frogs, *Theloderma* spp., from China and Southeast Asia, attach their eggs to a suitable substrate, either as an egg-mass or singly, and when they hatch from 10 to 20 days later they slide or drop down into the water. Other species using this method include one of the American leaf-frogs, *Cruziohyla craspedopus* (*Phyllomedusa craspedopus*), and a microhylid *Ramanella montana* from India.

Similarly, the green-backed mantella, *M. laevigata*, from Madagascar uses tree-holes and broken bamboo in which water has collected, laying their large white eggs on the inside wall of the cavity. The tadpoles feed on eggs of other frogs that breed in the same tree-hole, on fertilized eggs of their own species, or on unfertilized eggs laid specially by the female. This closely parallels the life-cycle of several poison dart frogs belonging to the genus *Oophaga*.

BELOW LEFT Some reed frogs. *Hyperolius* spp., attach their eggs to leaves or branches overhanging water.

BELOW RIGHT Well-developed eggs of the mossy frog, *Theloderma corticale*, suspended above water in a cavity in a tree.

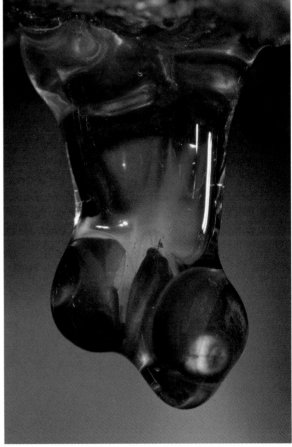

Laying eggs on land and carrying tadpoles to water

In another variation, tadpoles hatching from eggs that are laid on land may be carried to water by one or other of the parents. This involves a high degree of parental care because the adult has to remain with the eggs until they hatch, which can take several days or even weeks. Tadpoles are transported from a terrestrial laying site to water in the Seychelles frog, *Sooglossus sechellensis*, and in two guardian frogs from Borneo, *Limnonectes finchi* and *L. palavanensis*, in which males stay with the eggs, laid under a leaf on the forest floor, and move them to a small pool when they hatch.

The most studied species in this category, however, are the poison dart frogs, Dendrobatidae. Parental care is highly developed in these species, which lay small clutches of eggs on a leaf that the male has carefully cleaned by wiping or flicking movements with his feet. One or other of the parents, or sometimes both, depending on the species, attend the eggs daily, moistening them with bladder water if necessary, until they hatch. The newly hatched tadpoles wriggle onto the parent's back where they adhere quite strongly. In some species the parent takes them to a stream to continue their development but in others they are taken to a small body of water, often in a bromeliad plant. Again depending on the species, the tadpoles may be carried en masse or individually. In members of the genus *Oophaga*, such as the strawberry poison dart

ABOVE A male rough guardian frog, *Limnonectes finchi*, carrying his tadpoles in a temporary pool in the Danum Valley, Borneo.

LEFT Poison dart frogs lay small clutches of eggs on land and transport them to water when they hatch.

RIGHT The strawberry poison dart frog, *Oophaga pumilio*, and many other dendrobatids make use of the pools of water that collect in bromeliad plants to house their tadpoles.

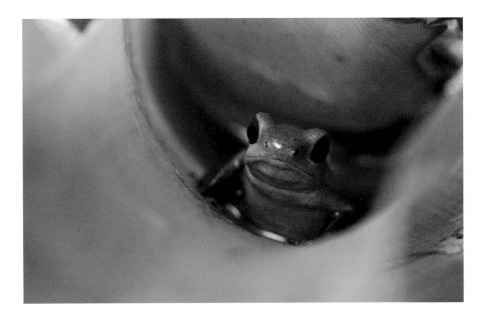

RIGHT The strawberry poison dart frog, *Oophaga pumilio*, and many other dendrobatids make use of the pools of water that collect in bromeliad plants to house their tadpoles.

RIGHT A male three-striped poison dart frog, *Epipedobates trivittatus*, from Surinam, transporting a large number of tadpoles to water.

frog, *O. pumilio*, the females lay their small clutches of two to six eggs on the forest floor and stay near them. When they hatch she shuffles into the jelly mass and encourages one tadpole at a time to climb onto her back. She then climbs into a tree and finds a suitable deposition site, such as a bromeliad plant, and releases the tadpole into the water, before returning to the clutch for another tadpole. She does this for each tadpole until the entire clutch has been transferred. Remembering where she put them, the female visits each tadpole-holding bromeliad every day or two and lays one or more infertile eggs into the water for the tadpole to eat.

Carrying eggs or young, or both

The vulnerability of the egg stage is overcome in some species by carrying the clutch until they have hatched into tadpoles. In the marsupial frogs, *Gastrotheca* spp., and several related genera in the family Hemiphractidae, all from Central and South America, the female carries the eggs on her back, sometimes exposed, as in *Hemiphractus*, and sometimes in a pouch, as in *Flectonotus* and *Gastrotheca*. The male fertilizes the eggs as they are laid and then manoeuvres them onto the female's back or into her pouch. The Riobamba marsupial frog, *G. riobambae*, for example, can carry up to 100 eggs and uses the long toes of her hind feet to scoop out the tadpoles when they have hatched. Other species of *Gastrotheca* and some of the other members of the same family, carry their eggs and tadpoles until they are fully developed, releasing them as small, fully metamorphosed, frogs.

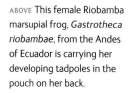

The totally aquatic pipa toads all carry their eggs on their backs although the stage at which they leave differs with the species. The eggs are positioned on the female's back during a complicated spawning sequence in which the paired toads swim up to the surface of the water and turn upside down. A few eggs are laid at the top of each loop and the male carefully manoeuvres them onto the female's back using his large webbed hind feet. This is repeated until all the eggs are laid, usually around 100. During the next 24 hours a pad of spongy tissue swells around the eggs until each one is almost completely embedded in its own cell, where it develops. In the Surinam toad, *Pipa pipa*, the larvae remain in the female's back until they are fully formed toadlets, whereas in other species they are released as tadpoles.

ABOVE This female Riobamba marsupial frog, *Gastrotheca riobambae*, from the Andes of Ecuador is carrying her developing tadpoles in the pouch on her back.

LEFT A pair of dwarf pipa toads, *Pipa parva*, spawning. The male is gathering the eggs as they are laid and using his large hind feet to manoeuvre them onto the female's back.

Midwife toads, *Alytes* spp., which occur in western Europe and a small area in North Africa, differ from the previous species because the male carries the eggs, wrapping them around his hind legs as they are laid. A clutch can total up to 80 eggs, although they usually number less than half this. Some males carry more than one string at different stages of development, due to having mated with two or more females. He keeps the eggs moist by staying in a damp burrow, often under a rock, and may visit a pond to re-hydrate them during dry weather. The eggs take about a month to hatch at which point he releases them into a small body of water.

ABOVE A male European midwife toad, *Alytes obstetricans*, from the Picos de Europa, Spain, carrying a string of eggs.

In Patagonia, Darwin's frog, *Rhinoderma darwinii*, lays its eggs on damp ground and, after fertilizing them, the male picks them up in his mouth and manoeuvres them into his vocal sac. Here they develop, nourished partly by their yolk and partly by secretions from the frog's skin for about 52 days, by which time they are fully developed and the male spits them out. The only other member of the genus, *R. rufum*, also carries its eggs but this species deposits the tadpoles in water to complete their development.

The Australian marsupial frog, *Assa darlingtoni*, which is not related to the American species with the same common name, also has a unique breeding method. The eggs are laid on the land and the male stays nearby. When they hatch, about 11 days later, he enters the jelly mass and the tadpoles move through the jelly layer and enter one of the two small pouches situated on his hips. They continue their development in the pouches for 48–69 days before emerging as young froglets.

RIGHT Darwin's frog, *Rhinoderma darwinii*, with newly emerged young.

Even more remarkable are the gastric brooding frogs, *Rheobatrachus* spp., also from Australia. Shortly after the first species, *R. silus* was discovered in 1973 it was shown that the female swallows up to 26 of her eggs at some point after they have been laid and fertilized; spawning was never observed. Her digestive juices are switched off so she does not digest the tadpoles as they develop in her stomach, which eventually becomes so distended that it fills almost the whole of her body cavity. Breeding was never observed in the wild but, in the laboratory, development took six weeks and the fully formed young were disgorged by the female over a period of several days. Tragically, the species has not been seen since 1981, and extensive searches have failed to locate it. A second species of gastric-brooding frog, *R. vitellinus*, was discovered in 1984, but this was last seen in 1985. Both species are presumed to be extinct.

Direct development

In this method of reproduction frogs have completely broken the link with free-standing water. Eggs are laid on the ground in depressions in the soil, in leaf-litter, among moss, or under rocks or logs, and the entire developmental process takes place within the egg capsule. Species with direct development tend to be small and to lay few eggs, although some of them lay at frequent intervals throughout a large part of the year so the total number of eggs may be high. The most numerous frogs of this type are the members of the Eleutherodactylidae and the Microhylidae.

Nearly 200 species of litter and rain frogs, *Eleutherodactylus* spp., and the members of several related genera, dominate many frog communities in Central and South America, and the West Indies, due to their independence of water, short time to reach maturity and high reproductive potential. Females typically lay clutches of about 20 eggs but larger species are capable of laying up to 100. The eggs are relatively large and stuck together in a cluster, and are usually hidden under dead leaves or other forest debris, although some arboreal species use tree-holes, bromeliads and other sites. The tadpole, developing inside the egg capsule, has no mouthparts and is a fully developed froglet when it hatches. In at least one species, *E. cundalli*, from Jamaica, the female stays with the eggs until they hatch and the young climb onto her back.

In New Guinea, Australia, and neighbouring islands, a parallel situation exists within the Microhylidae, specifically those in the subfamily Asterophryinae, of which there are 243 species. Although the natural history of some is unknown, those species that have been observed all have direct development. In many (perhaps all) the male stays near the eggs, often in contact with them, while they are incubating, which can take up to three months. Males in several genera, including *Sphenophryne* and *Cophixalus* spp., which are sometimes known as nursery frogs, have been found with froglets clinging to their backs. They jump off the male as he moves about and the purpose of transporting them may be to distribute them over a larger area and so avoid competition with each other. Staying in the New Guinea region, 84 species in the Ceratobatrachidae (formerly part of the Ranidae) also have direct development. These include the many *Platymantis*

RIGHT A newly hatched nursery frog, *Cophixalus ornatus*, still with remnants of its tail. Males of this species carry their young away from the egg-laying site.

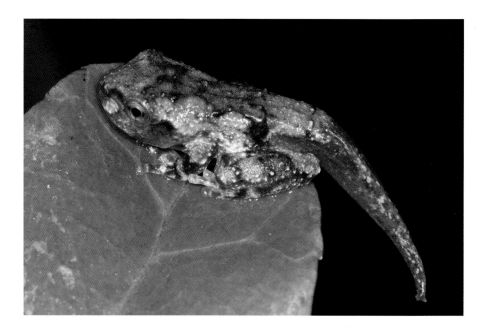

species, some of which are arboreal, and the unusual Solomons leaf frog, or triangle frog, *Ceratobatrachus guentheri*.

Apart from these two major groups of terrestrial-breeding frogs there are others, from a variety of families around the world. They include several small bufonid toads from South America and Africa, Gardiner's Seychelles frog, *Sechellophryne gardineri* (*Sooglossus gardineri*), Asian tree frogs in the genera *Philautus* and *Rhacophorus*, and several burrowing toads from Australia. In the two sandhill frogs of the *Arenophryne* genus for example, mating takes place underground and the 6–11 eggs are laid in burrows up to 80 cm (31 ½ in) deep. Under laboratory conditions the eggs hatch in 10 weeks. The turtle frog, *Myobatrachus gouldii*, from the same region and in the same family, Myobatrachidae, has a similar breeding system but digs down even deeper, up to 1 m (40 in), to lay its eggs.

Birth to live young

Independence of water reaches its peak in a small number of frogs that give birth to live young. Internal fertilization is an obvious requirement although not all frogs with internal fertilization are live-bearers, the tailed frog being an example of a species that has internal fertilization but lays eggs. There are two forms of live-bearing, and both occur in frogs. Where the frog retains the eggs in her oviduct but does not contribute nourishment then this is ovo-viviparity, whereas when the developing tadpole is nourished in some way by the mother this is termed viviparity. The golden coqui, from Puerto Rico, *Eleutherodactylus jasperi*, is an example of the former. It only produces three to five eggs and they are very large in relation to its body size. Development takes about one month and the developing young are nourished by their yolks in the same way as terrestrial-breeding *Eleutherodactylus* spp. *Eleutherodactylus jasperi* is probably extinct but, because the life-cycles of many species are unknown, it is quite possible that other members of the genus are ovo-viviparous.

A small group of African toads includes species that have a tendency towards live-bearing. This group contains species with internal fertilization but which lay their eggs at an early stage of development, species that retain their eggs until they hatch (*Nectophrynoides tornieri* and *N. viviparus*) and a species that nourishes its young with a substance called uterine milk while they are in the uterus. This last species is the Nimba toad, *Nimbaphrynoides occidentalis*, from Liberia, Ivory Coast and Guinea. (It is sometimes considered to be two separate species, *N. liberiensis* and *N. occidentalis*, and, to confuse matters even more, they were previously placed in the genus *Nectophrynoides*.) This species gives birth to 4–35 young, and gestation can take as long as 270 days, but this is partly due to the long period of aestivation it sometimes has to spend underground during its pregnancy to escape dry conditions.

Parental care

A degree of parental care is present in a number of the reproductive modes described above. Parental care of some kind has been recorded in about half of all frog families and estimates of the total number of frog species that care for their eggs and/or young range from 6 to 15%. Given the lack of information about so many species, and bearing in mind that the existence of parental care in many species has only been discovered recently, this figure could be much higher.

Parental care takes many forms. Simply staying close to the eggs, as in the glass frogs and some microhylids, can help to deter predators and possibly parasites and fungal infections. A step further involves helping the tadpoles to a suitable place to continue their development, as in the shovel-nosed frogs that break through the wall of their underground chamber to lead their tadpoles to an adjacent stream, or using their bodies to bulldoze a route from a drying pool to a larger one in the case of the African bullfrog. Other species physically transport their tadpoles from the place where the eggs were laid to a more suitable place for their continued development, as in many poison dart frogs and the guardian frogs, or it may go as far as carrying them for all of their subsequent development, as in some American marsupial frogs and the Surinam toad. Others, including the green-backed mantella and several tree frogs, lay infertile eggs for their tree-hole-dwelling tadpoles to feed on, while a group of poison dart frogs deposit their tadpoles individually and return at regular intervals to feed them on infertile eggs, showing a remarkable ability to remember where they are and to anticipate their needs. Several species, such as the gastric-brooding frogs and Darwin's frog (see p.102) have evolved unique systems for caring for their young. These species carry their developing young in the stomach and vocal sac, respectively. A select group of frogs, however, carry them in their uterus. It is probably true to say that frogs show the greatest diversity of reproductive methods of any vertebrate group.

8 Habitat and distribution

THE DISTRIBUTION PATTERN OF FROGS AND TOADS around the world is the result of a number of factors, particularly the suitability of the habitat in terms of food, shelter, food and breeding sites, and its accessibility. Even within relatively small areas there will be differences between the number of species and the density of populations from one place to another. These differences are due to variations in the quality of the habitat in so far as it affects frogs, and the availability of the area to expanding populations of frogs. A perfectly good habitat may have no frogs if it is inaccessible. Barriers to dispersal include oceans, deserts and mountain ranges, while dispersal corridors include valleys and forests. With the changing shape of the landmasses and periods of warming and cooling, present corridors and barriers are not necessarily the ones that have created the present distribution of frogs.

OPPOSITE Flying frogs, such as the harlequin flying frog, *Rhacophorus pardalis*, are most common in forests that have a fairly open structure.

Habitats

Although habitats are continuous, with one often merging into another, it is convenient to divide them roughly into types. Some frogs are very specialized and only found in one kind of habitat, whereas others are more generalized, and found in many different habitats. Generalist species are those that are able to survive in a variety of habitats and sometimes move from one to another, when they migrate from terrestrial habitats to ponds in the breeding season, for example. Others live at the interface of aquatic and terrestrial habitats their whole lives by inhabiting the margins of pools, lakes and rivers, and yet others may be terrestrial but climb into low bushes to feed, call or hide. Most types of habitat seem to produce a few specialists, however, and these are of interest because they show how frogs' shapes and behaviours are moulded by the places in which they live.

Aquatic frogs

Although many frogs live in or around aquatic habitats, few species are completely aquatic. All 32 members of the tongueless frogs, Pipidae (seven in South America and the rest in Africa) are firmly tied to water and cannot feed on land, although they sometimes

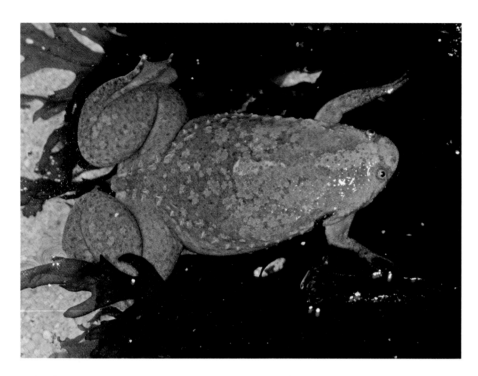

OPPOSITE AND LEFT The African clawed frog, or common platanna, *Xenopus laevis*, is an adaptable and wide-ranging species that lives in a variety of aquatic habitats including ponds such as this in Namibia.

BELOW The eyes and nostrils of semi-aquatic frogs such as Budgett's frog, *Lepidobatrachus laevis*, are positioned on top of their head so that they can see and breathe while most of their body is submerged.

make their way across it to colonize new ponds and lakes. They have powerful hind limbs and large, fully webbed hind feet for swimming. Their eyes are small and they rely more on other senses, such as lateral line organs, to navigate through the murky water in which they live and find prey. The *Pipa* species also have small star-shaped appendages on the tips of their front digits, which are sensitive to touch. Apart from four species belonging to the genus *Hymenochirus*, they do not have webbed front feet, but use their front limbs like forks, to manoeuvre prey into their mouths, compensating for their lack of tongues.

Other aquatic frogs, such as the paradoxical frogs, *Pseudis* spp., the Andean lake-dwelling frogs belonging to the genus, *Telmatobius*, and the Chilean water frogs, also have large fully webbed hind feet but lack the specialized sense organs. Semi-aquatic frogs, a category that covers most species, have webbed hind feet and their eyes and nostrils are positioned towards the top of their head so that they can hang just below the surface with only their eyes showing.

Torrents and waterfalls present frogs with a number of challenges as well as an opportunity to exploit a new niche. The ghost frogs, *Heleophryne*, from South Africa, the tailed frogs, *Ascaphus*, from the American Northwest, various torrent frogs, *Meristogenys* and *Staurois*, from Southeast Asia, and *Taudactylus* from Australia, are all specialists, not found in other types of habitat. These genera, and others like them, live in the splash zones of waterfalls, in rock crevices behind and alongside them, or on emergent rocks in fast-flowing cascades. Adaptations include discs on their toes to cling to wet, slippery rocks, and a low profile, to reduce water resistance. Their tadpoles invariably have some means of attaching

themselves to the substrate, usually in the form of a large sucker-like disc surrounding their mouth, to prevent them being washed downstream. Stream frogs, such as the *Staurois* spp., communicate visually by signalling with the hind feet, the webbing of which is brightly coloured, in a form of semaphore. The Panamanian golden frog, *Atelopus zeteki*, which lives along fast-flowing streams, has a similar behaviour, although it uses its front limbs and Australian torrent frogs, *Taudactylus* spp., also display by arm and leg waving.

Burrowing frogs

Frogs that spend most of their lives underground often do

TOP A rainforest stream on Mount Kinabalu, Borneo is a rich habitat that is home to many species of frogs.

ABOVE The black-spotted rock frog, *Staurois natator*, lives and breeds along the margins of fast-flowing streams.

so because they live in a hostile environment, such as a desert, and need to avoid drought and extremes of temperature, or they may simply be exploiting an underused niche. Frogs that live in deserts, or other places that experience periods of dry weather, can often be recognized by their wide heads, short limbs and rotund bodies, a shape that helps to reduce evaporation, and they often have adaptations that enable them to burrow more efficiently such as the hardened crescent-shaped tubercle on the undersides of their hind feet, which has given some of them the common name of spadefoot toads. These toads, and other species with a digging tubercle, burrow backwards by shuffling their hind feet into sandy or friable soil. As they go down they tend to rotate slightly so they appear to be twisting themselves into the ground. A number of species can remain underground for extended periods by forming a cocoon of a thickened layer of epidermal skin around their bodies, which reduces evaporation considerably. Burrowers of this type come to the surface in large numbers to breed synchronously after rain, although they often differ in the amount of rain needed to stimulate them, and their tadpoles often grow and develop very quickly (see p.92 and p.141).

BELOW The Karoo toad, *Vandijkophrynus gariepensis*, (formerly *Bufo gariepensis*), from South Africa is a desert-adapted toad.

Other frogs burrow head-first, and these include the African shovel-nosed frogs, *Hemisus* spp., which form the family Hemisotidae, two species of sandhill frogs, *Arenophryne rotunda* and *A. xiphorhyncha*, and the turtle frog, *Myobatrachus gouldii*, all from Australia. These frogs have powerful forelimbs for pushing their way through soil and modifications to their skulls or neck vertebrae to give them greater thrust. Their snouts are often pointed and they start burrowing by tilting their head down and diving into the soil. As they progress they use their front feet to pull themselves down and they may also use their hind feet to push surplus soil out of the burrow. Finally, a number of head-first burrowing microhylid frogs spend most of their lives underground, forcing their way through leaf-litter, between matted roots, and under rocks. All these species feed beneath the surface and some of them breed there too, laying their eggs in underground chambers where they undergo direct development. Many of them are only seen by chance when soil is turned over in the course of agriculture or other groundwork and it seems quite likely that many more frogs of this type await discovery.

BELOW Even the most unlikely looking habitats are populated by frogs. The Namaqualand desert of South Africa is home to several burrowing species.

Arboreal frogs

Arboreal habitats are most numerous in tropical regions where rainforests cover large areas in South and Central America, Africa, Southeast Asia and eastern Australia. Forests have different characteristics according to where they are and the mix of tree species that grow, some have dense understorey vegetation whereas others are gallery forests with open space between ground level and the canopy. Growth is often most dense along the edges of rivers and streams, so frog populations here can be very rich in species and in numbers. There are frogs at all levels and many species live at specific heights, sometimes moving to different levels at different times of the day and night.

Arboreal frogs are typically slender-bodied, and light in weight, with long limbs to give them greater leverage, and expanded toe discs for clinging and climbing. When resting on vertical surfaces for extended lengths of time, many arboreal frogs bring their undersides into contact with the surface, to increase the adhesive area. A number of species in two separate families, the Hylidae and the Rhacophoridae, have heavily webbed feet for gliding (see p.23). These species rely on open forest and it is no coincidence that other flying animals, such as flying lizards, snakes and squirrels, share their habitat. Within the Hylidae, the so-called monkey frogs of the genus *Phyllomedusa* have opposable toes for grasping thin twigs and branches and walk with a deliberate, stiff-legged gait, like chameleons. Arboreal frogs rarely live in dry environments, not only because there are fewer trees there, but also because it is more difficult for them to escape the drying effects of moving air. The few tree frogs that do live in dry regions, such as the waxy frog, *Phyllomedusa sauvagii*, from the Chacoan region of South America, and the African grey tree frogs, *Chiromantis* spp., from East African savannas, secrete a waterproof substance from their skin to reduce evaporation.

Arboreal frogs are found in many families other than the tree frog families, Hylidae and Rhacophoridae. They occur among the glass frogs, poison dart frogs, marsupial frogs,

BELOW Rainforest canopies provide habitats for arboreal species many of which rarely, if ever, descend to ground level.

LEFT An important characteristic of New World rainforests is the presence of many bromeliad plants, which hold water in their leaf axils and provide arboreal frogs with places to hide and breed.

mantellas, ranids and microhylids and even in such unlikely families as the Bufonidae, where the Asian tree toads, *Pedostibes* spp., live in large forest trees, and climb down to ground level to breed. Others breed in the canopy, using pools of water that form in the centres of plants or in leaf axils to lay their eggs, or to deposit their tadpoles. Several species breed in tree holes, and South and Central American species benefit from the abundance of water-holding bromeliad plants in which many of them live and breed, never needing to come down to ground level their entire lives. A new poison dart frog, *Oophaga arborea* (originally described as *Dendrobates arboreus*) from Panama, was only discovered in 1984 when large trees were felled during the construction of a road, giving scientists access to their bromeliad-festooned crowns.

Distribution

There are few studies on the distances individual frogs and toads cover during their lifetimes. Foraging for food and migrating to and from breeding sites are the most important motivating factors so there are differences between species that live amongst their food (burrowing frogs that live in or near termite colonies, for example) and those that have to forage widely for it. Similarly, some frogs can breed almost anywhere – if they lay their eggs in leaf-litter or moss, for example – whereas others make heroic journeys to breeding ponds each spring. The European tree frog, *Hyla arborea*, measuring a mere 4–5 cm (1 ½–2 in) has been known to travel over 12 km (7 ½ miles) in a year and the European common toad, *Bufo bufo*, may migrate up to 3 km (1 ¾ miles) to get to its breeding pond, sometimes crossing major roads in the process.

Leaving aside breeding migrations, most individuals seem to live in a fairly well-defined area, or home range, but the difficulties in monitoring often means that generalizations of this kind are based on anecdotal evidence. Some frogs and toads can be seen in the same place day after day (see p.69) and it seems likely that, once they find a suitable environment and a reliable food supply, they move very little. The extent of the area covered in the course of their movement over a period of time, known as their home ranges, for the few species that have been studied include an average 2,117 sq m (22,787 sq ft) but this can be misleading. Striped chorus frogs, *Pseudacris triseriata*, for instance, can have ranges of over 6,000 sq km (2,317 sq miles) although this figure includes breeding migrations. On the other hand, Johnstone's whistling frog, *Eleutherodactylus johnstonei*, a species that has direct development and lays its eggs in leaf-litter, was found to stray less than 2 m (6 ½ ft) from their original capture site over a 300-day study period.

Global distribution

The evolution of frogs coincided with momentous events in the world's history, with landmasses breaking apart, drifting one way and then another, and colliding again, sometimes forming new continents. Because saltwater forms a barrier to frogs' dispersal, the only way they can reach new continents is to evolve there, migrate there across land bridges, or more recently by introduction into a new region. Superimposed on this pattern is one of changing climatic conditions, competition and the expansion and contraction of populations as the environment changed. Some families became trapped in small regions, prevented from spreading by barriers such as mountains and wide rivers while others made sweeping advances across wide tracts of suitable habitat. Some families became extinct, others lingered on but contain only limited numbers of surviving species. What we have left is a complex pattern of distribution in which some continents have more families and species while others, apparently just as suitable for frogs, have few. The following survey is not intended to be comprehensive but gives a brief overview of the way in which frogs from different families are distributed around the world.

North America (the Nearctic region) has relatively few families and species of frogs despite its large size. It was separated from South America until about three million years ago but joined to Eurasia, so in many respects its frog fauna has more in common with that region than with its southern neighbour. It has large numbers of toads (Bufonidae), tree frogs (Hylidae) and true frogs (Ranidae). The single species within the Rhinophrynidae (the Mexican burrowing frog) is found only here, as are all seven of the American spadefoot toads (Scaphiopodidae). Two tailed frogs, *Ascaphus* spp., are relics of the most ancient frog family, the Leiopelmatidae, with the other four species occurring in New Zealand. Otherwise, only small numbers of mainly Central and South American families just reach the continent and one introduced member of the Pipidae (*Xenopus laevis*, the African clawed frog) occurs here.

South and Central America (the Neotropical region) with much of it lying within the tropics and covered with rainforest, is rich in frogs. Several families are endemic to the region or almost so, including the poison dart frogs (Dendrobatidae and Aromobatidae), the glass frogs (Centrolenidae), the marsupial and backpack frogs (Hemiphractidae) and a number of families formerly included within the Leptodactylidae: Calyptocephalellidae,

Ceratophryidae, Cycloramphidae, Eleutherodactylidae and Strabomantidae. One branch of the Pipidae lives here, the other being in Africa, and there are large numbers of tree frogs (Hylidae). The otherwise widespread Ranidae, however, is represented only by two species in the north of the continent, these having moved down from North America in more recent times, made possible by the relatively recent formation of the isthmus of Panama.

Europe, northern Asia and Mediterranean Africa (the Palaearctic region) is relatively poor in frog species and families. The spadefoot toads (Pelobatidae), parsley frogs (Pelodytidae) and the midwife toads (Alytidae) are endemic to the region but all are small families. Several of the fire-bellied toads, Bombinatoridae, occur here and there are three species of tree frogs, several toads and a number of water frogs. The dominant families are the Bufonidae and Ranidae.

Africa, excluding the Mediterranean coast (the Ethiopian region) is rich in endemic families, including the squeakers and bush frogs (Arthroleptidae), rain frogs (Brevicipitidae), ghost frogs (Heleophrynidae), the shovel-nosed frogs (Hemisotidae), reed frogs (Hyperoliidae), the African water frogs (Petropedetidae), the puddle frogs (Phrynobatrachidae), the grass frogs (Ptychadenidae) and the pyxie frogs, (Pyxicephalidae). Also present in good numbers are the Bufonidae and Ranidae. The Microhylidae is well represented on Madagascar, but less so on the African mainland, while the Mantellidae is endemic to Madagascar. Madagascar is rich in species and nearly 400 have been described so far, all of them endemic.

South and Southeast Asia (the Oriental region) is another rich region with several endemic families, including the horned and litter frogs (Megophryidae), the Micrixalidae and the Ranixalidae, neither of which have collective common names. The Indian purple frog, *Nasikabatrachus sahyadrensis*, is the only member of its family (Nasikabatrachidae) and the Seychelles frogs (Sooglossidae) are restricted to the islands of the same name. The Asian tree frogs (Rhacophoridae) are almost endemic with just a few species in Africa. Other well-represented families include the toads (Bufonidae), water frogs (Ranidae), and the Microhylidae. There are no members of the Hylidae, their niche being filled by the Rhacophoridae.

Australia, New Zealand and New Guinea (Australasia) has been separated from other landmasses for a long time, perhaps for 40–80 million years, and it therefore has several endemic families. The Australian ground frogs (Limnodynastidae) and the froglets and toadlets (Myobatrachidae) are endemic to the region and the Microhylidae are well represented, especially in New Guinea and surrounding islands. The only frog family native to New Zealand is the Leiopelmatidae, with four of the six species, the other two being the tailed frogs, from North America. The Ceratobatrachidae, as presently understood, is nearly endemic, with just one species, *Ingerana baluensis*, from Borneo, being found outside the region (although this species may be moved to another family in the future). Many species in this family are only found on the Solomon Islands. There are many water frogs, Ranidae, although only one species reaches Australia and only in the extreme northeast. Tree frogs (Hylidae) are well represented, with many species of *Litoria* in Australia and New Guinea. Toads (Bufonidae) are absent from the region apart from two species introduced to New Guinea, one of which, the cane toad, has also been introduced to Australia.

9 Frogs and man

IN THE LONG AND CHEQUERED HISTORY of frogs' associations with humans, they have figured in religion, folklore, witchcraft and children's literature, and provided food, biological control, poisons and medicines. Many students of biology will have dissected a frog, and until the 1960s frogs were used in pregnancy tests. More recently, frogs around the world have suffered at the hands of humans, and their numbers are shrinking at an alarming rate. The causes are varied and in many cases it seems that there is an interplay between two or more factors. An important part of current research is directed towards ensuring that frogs have a future.

ABOVE Carved turquoise frog charm or fetish made by North American Zuni Indians.

Uses

Chinese myths concerning toads are among the oldest. The Chinese believed that the world rested on a giant frog and that lunar eclipses occurred when a frog tried to swallow the moon. Another character, Ch'an Chu, is a three-legged frog, also known as the 'lucky money frog', a symbol of fruitfulness and prosperity. Figures of this character, sitting on a pile of Chinese coins and with a coin in its mouth, are believed to increase income and protect wealth. The Egyptians also incorporated frogs into their beliefs, and frog motifs are found on ancient pottery, metalwork and weaving. They used a hieroglyphic symbol of a tadpole to represent the number 100,000. Moses brought a plague of frogs down on the city of Rameses in revenge for the Pharaoh's refusal to release the Israelites even though the Egyptians worshipped frogs, in the form of one of their gods, Heket, who had the head of a frog and was the protector of newborns. Frog effigies are common in ancient Central and South American cultures such as those of the Aztecs and the Mayans, and stylized designs of them were frequently painted on the undersides of pottery vessels, an allusion to the fact that frogs and toads are often found underneath objects. It has been suggested that the ability of the secretions from toads' skin to induce hallucinogenic states when swallowed may have given them special significance, along with plant extracts such as those from the peyote cactus, that have the same effect. The Mayans also associated frogs with the god Chac, symbolizing water, crops, fertility and birth.

OPPOSITE A cane toad or marine toad, *Rhinella marina* (formerly *Bufo marinus*) in a sugarcane field near Gordonvale, Queensland, Australia, the town where it was originally introduced in 1935.

In medieval Europe, frogs and toads were synonymous with evil and witchcraft and toads are depicted as the familiars and alter-egos of witches. Almost all writings of the time portray the toad in a negative fashion (frogs are hardly mentioned) and Shakespeare refers to them as 'ugly and venomous'. Even today, toads are not welcome in some places and are thought to cause warts, while superstitions concerning toads were commonplace in parts of England and North America until recently. Kenneth Grahame's *The Wind in the Willows*, published in 1908, characterizes Toad as pompous and unreliable, but he redresses the balance by also describing him as intelligent and witty. In other children's tales from Europe, frogs are princes in disguise waiting only for a maiden's kiss.

Poison dart frogs and South American Indians

South American Indians of the Emberá Chocó group of tribes from the Pacific slopes of the Andes in Colombia use at least three species of poison dart frogs to tip their blowgun darts: *Phyllobates aurotaenia*, *P. bicolor* and *P. terribilis*. The poison consists of steroidal alkaloids in the frogs' skin secretions, of which the most powerful is produced by the golden poison dart frog, *P. terribilis*. Its secretions are 20 times as toxic as those from any other species of poison dart frog making it dangerous to handle – even contact with the skin can result in a painful burning sensation through any small scratch, and the Indians protect their hands with leaves when handling them. Experimentally, the poison from a single frog was enough to kill at least 20,000 mice when injected under the skin. Extrapolating to humans this is equivalent to six or seven average-sized people though comparing toxicity between mice and humans is not always completely accurate.

The Indians make their blowguns from a length of straightened palm wood that is split down the middle. Then they make a semicircular groove in each half before whipping them back together. The darts are made from hard slivers of palm wood about 21–23 cm (8 ¼–9 in) long. A fine spiral groove is cut in the final 2–3 cm (⅘–1 1/16 in) of the point and this holds the poison. The darts are charged with poison in two ways. For *P. aurotaenia* and *P. bicolor*, the frogs are impaled on sticks and may be held over a fire. This causes them to secrete enough toxin from their skin for up to 20 or 30 darts, which are then pushed into a piece of plantain to dry. In the case of *P. terribilis*, however, the darts are simply wiped over the skin of a living frog; they are collected and kept in small wicker baskets. When they need the frogs, they pin them down using short sticks. Each frog, known locally as 'kokoi', is used to charge two or three darts before being allowed to hop away. This species is apparently abundant and replacing them is relatively easy. The Indians use the blowguns and darts mainly for hunting but occasionally for fighting.

LEFT The world's most poisonous animal is the golden poison dart frog, *Phyllobates terribilis*. It is used by native people in parts of Colombia to tip their blowgun darts.

Frogs and medicine

In 1974 Professor John Daly of the National Institutes of Health collected skin samples from the phantasmal poison dart frog, *Epipedobates tricolor* for biochemical analysis. (As a matter of interest, Professor Daly was also the first to investigate the highly toxic skin secretions of the golden poison dart frog, *Phyllobates terribilis*.) Extracts from the skin samples were tested on mice, where they were found to be an effective pain-killer, and the active constituent was named epidatidine, after the genus to which the frog belonged. Epidatidine was found to be 200 times more effective than morphine and, furthermore, it was non-addictive. So far so good, but the quantity of the sample extracts was insufficient to explore the implications fully and the frog species had already disappeared from one of its two known sites in Ecuador by 1976, because its habitat had been turned into banana plantations. In addition, the species was now protected by CITES, preventing collection of large numbers from the remaining site. Captive-bred *E. tricolor* were readily available but their skin did not contain epidatidine, due to the fact that their diet was different from wild populations: free-living poison dart frogs sequester chemicals from the insects they eat – mainly ants – to use in the manufacture of their skin toxins.

Fortunately, Professor Daly had frozen a minute quantity of his sample and by 1990 technological advances allowed analysis of even this tiny amount. Two more years of research resulted in a breakdown of the compound into its component parts and, shortly after this, a synthetic version of epidatidine was produced. Although this was too toxic for use in humans, the additional material was now available for subsequent research and in 1998 the Chicago-based company Abbott Laboratories created a non-toxic, non-addictive pain-killer, without serious side-effects, derived from Epiditine, which was named ABT-594. Clinical tests of ABT-594 are ongoing and the outcome is hopeful.

ABOVE The phantasmal poison dart frog, *Epipedobates tricolor.*

As food

Frogs are widely eaten by humans. In tropical countries where frogs are common, local people catch and eat them in numbers that are probably sustainable, although the Goliath frog, which is considered to be a delicacy by the indigenous people of those parts of West Africa where it lives, may be declining as a result (although habitat destruction probably contributes more towards its plight). In parts of Europe, notably France, local consumption of frogs' legs has taken place for centuries, and the species most commonly eaten is the well-named edible frog, previously called *Rana esculenta* but now known to be an unusual type of hybrid between the pool frog, *Pelophylax lessonae*, and the marsh frog, *P. ridibundus*.

BELOW Dead frogs for sale in
a street market in Thailand.

BOTTOM The edible frog,
Pelophylax esculenta, is the
species eaten most in Europe.

In other parts of the world, frogs form a major part of the trade in meat. Local stocks often cannot satisfy demand, either because frogs have become rare, because they are protected, or because the demand rises too steeply for the supply to keep pace, and so the international trade in frogs makes up the shortfall. The major importers of frog meat are France and the United States, while the major exporters are Indonesia and China. Indonesia's exports of frog meat peaked at 5,600 tonnes in 1992, although its domestic market is estimated at between two and seven times this amount. This represents a huge number of frogs, and global trade in wild-caught frogs is thought to account for between 180 million and one billion frogs per year. The main species involved are the crab-eating frog, *Fejervarya cancrivora*, the brown mountain frog, *Limnonectes macrodon*, and the paddy frog, *Limnonectes limnocharis*. The American bullfrog, *Lithobates catesbeiana* (*Rana catesbeiana*) is also farmed in Indonesia for the meat trade and there have been attempts to farm it in other parts of the world.

LEFT The red-legged frog, *Rana aurora*, was widely hunted for food following the California gold rush. The subsequent release of the American bullfrog from failed frog farms has further affected its numbers.

BELOW A booklet published in 1932 by the American Frog Canning Company encouraging land-owners to set up frog farms in California following the decline in local species due to over-collecting. 'Ponds can be built in a day with dynamite' is among the more interesting pieces of advice given.

In the past, the Asian tiger frog, *Hoplobatrachus tigerinus*, was introduced to Madagascar as a source of meat and the American bullfrog was introduced to parts of the western United States to replace dwindling stocks of local species. In California, red-legged frogs were hunted from the beginning of the nineteenth century but demand soared following the California gold rush of 1849 to such an extent that the harvest had risen to 22,405 kg (49,394 lb) in 1889. By 1894, however, numbers had declined so much that only 10,000 kg (22,046 lb) could be harvested and so local land-owners in California's Central Valley were encouraged to start up frog farms to meet the demand, using bullfrogs imported from the eastern states. When these enterprises failed, the bullfrogs escaped, or were released, into local ponds and streams, where they still live. California was still importing frog meat, from foreign sources, until early 2010 when the California Fish and Game Commission approved a ban on imports of non-native frogs (and turtles) for the food trade.

Biological control

Frogs and toads are well-known as natural predators of invertebrates that can be pests of agriculture and are sometimes encouraged to live in gardens and greenhouses as a form of eco-friendly pest control. On a larger scale, at least three species of frogs, the green and black poison dart frog, *Dendrobates auratus*, the Japanese wrinkled frog, *Glandirana rugosa* (*Rana rugosa*) and the marine toad have been introduced into Hawaii as a form of biological control against pests of sugar cane, with mixed success.

The biological control programme that has made the most impact, however, is the introduction of the marine or cane toad, *Rhinella marina* (*Bufo marinus*) to Australia. This large species is native to Central America and northern South America, and adapts to a wide range of different conditions. It also eats large numbers of insects and breeds prolifically. The cane toad was introduced to many countries, some as far back as the mid-nineteenth century, in attempts to control pests of sugar cane, notably a beetle, *Phyllophaga vandinei*, often with some success. In 1934, after a successful introduction to Puerto Rico, cane toads were first taken to Hawaii and, from there, to the Philippines and neighbouring islands before being introduced to the Australian sugar-growing region centred on Gordonvale, Queensland, in 1935. Young toads bred from this initial stock were subsequently introduced to Cairns and Innisfail. The toads thrived, eating everything in their path, including beneficial insects as well as pest species. They had no competition and no important predators. By 1974 they had expanded their range until it covered approximately one-third of the land area of Queensland and by 1983 they had spread into the Northern Territory and New South Wales. Several episodes of accidental introduction followed, some as far away as Perth in Western Australia, although none of these resulted in established populations, as far as anyone knows.

The effect on Australian wildlife is not fully known. Several snakes, lizards, birds and mammals are known to have died as a result of eating cane toads but other species appear to be unaffected. The indiscriminate appetite of the toad probably means that it eats many other species, including other frogs, but firm evidence is lacking. They have been known to congregate around beehives and eat large numbers of bees and they certainly eat dog food and kitchen scraps put out for other animals. Fortunately, their breeding habits do not bring them into direct competition with other frogs because they breed in open water, a site used by only a small number of native species. They also seem to be concentrated in urban and agricultural areas, although it is by no means certain that they are not spreading into other types of habitat. Regardless of the ecological implications, coastal Queensland is overrun with cane toads, hopping about in gardens, fields and on the roads, where many meet an untimely death.

Pregnancy testing

In 1930, a South African scientist named Lancelot Hogben found that he could stimulate egg-laying in the African clawed frog, *Xenopus laevis*, by injecting them with hormones from cattle. Originally, the purpose was to provide amphibian eggs for embryological research but it soon provided doctors with a simple method of pregnancy testing in humans. When a female frog was injected with the urine from a pregnant woman, it laid eggs within 12 hours, confirming the pregnancy. This came to be known as the Hogben test. Due to its reliability and relatively low cost, the Hogben test was soon used in laboratories throughout Europe and the United States, which imported large numbers of the frogs from South Africa. By the 1960s, however, simpler methods of pregnancy testing had been developed and the frogs were no longer required, although many colonies were maintained so that the eggs and tadpoles could be used for other research projects. It seems that some of the frogs that were surplus to requirements were released into local ponds and waterways.

Declining amphibians

Until fairly recently, humans' interactions with frogs, such as those described above, have been limited to small numbers of species, in isolated places. The latter half of the twentieth century saw an acceleration in the rate of change to the environment and, as a result, the human impact on amphibian populations has increased. In addition to this, other factors that may be associated with human-induced changes have come into play, resulting in catastrophic declines in frog populations throughout the world. A number of causes have been proposed, some of them more significant than others. Factors are sometimes inter-related and so the causes cannot always be determined with any degree of certainty.

The present situation is exceedingly serious. The International Union for Conservation of Nature (IUCN) is the body that monitors populations of plants and animals and lists species under various grades from 'least concern' to 'extinct'. In 2002 they listed 157 species of amphibians as threatened with extinction but by 2008 this number had risen to 1,908, or approximately 30% of all amphibian species that they are aware of. Of these, 1,626 were frogs and they include 37 species that are probably extinct, 398 critically endangered, 650 endangered and 578 vulnerable. The most important reasons for amphibian declines are described below.

Habitat destruction and fragmentation

Habitat destruction has been implicated in many cases of declining amphibian populations. Logging, overgrazing, land drainage and land clearance for agriculture, commerce and residential development have all accelerated in the last 50 years as the human population has grown and expanded into new regions. This can happen on a large scale, when huge areas of rainforest are cleared to make way for oil palm plantations, or on a smaller, but cumulative, scale when settlers slash and burn native vegetation to create small plots to cultivate before moving on after a few years as the soil nutrients are used up or washed away. It has been estimated that 75% of the primary forest of Madagascar has been lost to this type of small-scale land clearance in the last 2,500 years, and Madagascar is by no means alone in this. Apart from the direct consequence of habitat loss when forests are removed, deforestation often leads to increased silting of rivers and streams,

RIGHT Pristine rainforest in the Ranomafana National Park, Madagascar. This is one of the richest places on earth for frog diversity.

CENTRE Outside the protection of a national park, a small paddy field has been created in a rainforest clearing.

BOTTOM The creation of more fields through slash and burn agriculture has led to total land clearance in central Madagascar. Very few frogs can survive in this environment.

making them unsuitable for certain types of tadpoles that require clear water and a gravel substrate in which to live. Planting of exotic trees, such as the conifers in the Cape region of South Africa can cause small streams to dry up altogether as they suck more moisture out of the ground.

Introduction of alien species

The introduction of animals to regions where they are not naturally found often has negative repercussions on local species but this can be difficult to assess. Introduced trout in high-elevation lakes of California and Nevada are implicated in the decline of the yellow-legged frogs, of which two species are now recognized and which have disappeared from over 90% of their former localities because the trout eat their tadpoles. Frogs themselves have been the subject of introductions, some of which are mentioned previously. The Cuban tree frog, *Osteopilus septentrionalis*, was introduced to Florida, probably in the nineteenth century. This is a large, adaptable species that dispersed throughout the state and eats other frogs. In the Everglades National Park it eats at least five other species of frogs and may eat additional species in other parts of the state. The African clawed frog was introduced and has become established in many American states as well as in Mexico, Chile, Europe and Southeast Asia, all of which have native frogs that could be negatively affected by the presence of a large and highly successful competitor. In the southwestern Cape region of South Africa, another clawed frog, *Xenopus gilli*, is endangered due to habitat changes and to hybridization with *X. laevis*, which is more adaptable.

ABOVE Introduction of the Cuban tree frog, *Osteopilus septentrionalis*, to North America occurred over 100 years ago. Since then it has had a negative impact on local species, which it eats and competes with for food.

Over-exploitation

Frogs and toads have been collected for the meat trade, for scientific research and teaching, and for the pet trade for many years. The demand for red-legged frogs in California and their subsequent decline is mentioned above, as is the collection of large numbers of Asian species for the meat trade. India banned the trade in frogs for meat as recently as 1987 because of concerns over insect pests, which were no longer being controlled naturally by the frogs. The European common frog and the American leopard frog were dissected in huge numbers by biology students up until the 1970s, and large numbers of frogs are imported to Europe and North America every year as part of the trade in exotic pets. The trade in plants and animals is regulated by the Convention on International Trade in Endangered Species (CITES). CITES lists 157 species of frogs in which trade is restricted and these include all species of poison dart frogs and all the mantellas, although some trade is allowed for small numbers of certain species. Species that might be most threatened by collection for the pet trade are those with limited ranges but, so far, there is no evidence that any frog species has been pushed to extinction through over-collecting. In some cases, the commercial value of frogs to a local community may be an incentive to preserve their habitats.

Pollution

Amphibians are particularly susceptible to pollution because they need clean air to breathe and, in most cases, unpolluted water for their eggs and tadpoles. Types of pollution that have been implicated in local declines include run-off of noxious chemicals from roads and fields, and acid rain. Spraying with insecticides also reduces their food supply. Contamination of water may stunt the growth of tadpoles or result in deformities, a phenomenon that has increased in the last two decades. Less obviously, it may affect hormone systems and the development of gonads, leading to infertility. In many cases we simply do not know what long-term effects chemicals in the environment are having on amphibians.

Climate change

The effects of climate change are difficult to assess. If warmer, drier conditions lead to droughts then frogs may be prevented from maintaining a suitable water balance, or they may die from heat exhaustion. If ponds dry up, or seasonal rainfall fails, then they are without breeding sites, or many more frogs may try to breed in the few remaining ponds, increasing competition. Raised levels of ultra-violet radiation, through the depletion of the ozone layer, may have harmful effects on the adults or their eggs and tadpoles. All these climatic effects, and others, have been proposed as possible causes for amphibian declines but they are difficult to test.

Infectious disease

Extinction of a species due to an infectious disease is almost unprecedented. In the normal course of events an epidemic kills off a proportion of a population but as the density is reduced the rate of infection slows down until it is no longer significant. Several types of viruses affect amphibians, the most important of which are ranaviruses. These viruses have caused epidemics in frogs in several parts of the world but they do not lead to extinctions because they are only transmitted when population densities are high; after an epidemic, population densities fall and the virus dies out, at least until numbers build up again. For a disease to cause extinctions it must be able to survive when the population of its host species is reduced to a low level. This only happens if it is fatal to some species but not to others, so that members of the more resistant species become sources of infection for those that are more susceptible.

The chytrid fungus epidemic

The scientific community first suspected that frogs were dying out in the 1980s. Several cases of sudden disappearances of frogs from places where they were formerly common were noticed and, worryingly, these were places where the habitat was protected and in good health in other respects. Furthermore, these extinctions occurred in widely separated places. They included litter frogs, *Eleutherodactylus* species, from Puerto Rico, stream-breeding frogs in Queensland, Australia, populations of western toads, *Anaxyrus boreas* (*Bufo boreas*), from the Rocky Mountains of Colorado, frogs and toads from the montane rainforests of Costa Rica and Panama, and montane frogs in South America,

from Venezuela to Peru. In the protected forests around Monteverde, in Costa Rica, for example, 20 species, about 40% of the total, had disappeared between the mid-1980s and 1994, including the golden toad, *Incilius periglenes* (*Bufo periglenes*). Numbers of other species, such as the harlequin toad, *Atelopus varius*, had also declined drastically. In Australia, both species of the gastric-brooding frogs, *Rheobatrachus silus* and *R. vitellinus*, disappeared, along with the Mount Glorious torrent frog, *Taudactylus diurnus*. Several other torrent frogs, two tree frogs, *Litoria* species, and a lace-lid frog, *Nyctimystes dayi*, had disappeared from the higher parts of their former ranges, although some of them could still be found at lower elevations. In the Andes, from Venezuela to Peru, frogs from montane habitats were also disappearing, most notably the small *Atelopus* toads that breed in mountain streams. Several of them became extinct in the late 1980s and practically all the surviving species are critically endangered. One survey of *Atelopus* in 2006 estimated that as many as 67% of the total number of species had disappeared.

BELOW Dead mountain yellow-legged frogs, *Rana muscosa*, floating in shallow water in Canyon National Park, California, 2006. They are victims of the chytrid fungus outbreak affecting frogs in many parts of the world.

When scientists working on amphibians around the world began to exchange notes on their experiences it became clear that the amphibian declines in pristine and widespread habitats were occurring much more quickly than they could explain in terms of habitat destruction, pollution or climate change even though these factors have undoubtedly had some effect on frog populations in some places, and may even have caused local extinctions. In the meantime, a fungal disease had been discovered in amphibians in the

early 1990s. It belongs to a group called the chytrid fungi and, although there are over 1,000 different forms of chytrid fungi, this was the only species known to parasitize a vertebrate host. In 1999 it was named *Batrachochytrium dendrobatidis*, or Bd for short. We now know that Bd only lives on amphibian skin, and often causes the death of frogs. Importantly, Bd infects *all* species of amphibians and so it remains a threat even after one or more species has disappeared from a region. The researchers reached the conclusion that Bd was the main cause of amphibian declines in pristine habitats and set out to investigate it.

There were several factors at work. Frogs at higher elevations are more likely to be affected than those at lower ones, indicating that the fungus is more virulent at the cooler temperatures experienced in mountains than the warmer conditions lower down. It seems that the optimal temperature for the fungus is 17–25°C (62.5–77°F) which is also, unfortunately, the temperature at which frog diversity tends to be greatest. Frogs restricted to cool mountains, especially stream-breeders, are the worst affected. Higher temperatures reduced the activity of the fungus and can even be used to cure infected frogs under laboratory conditions. Species that tend to aggregate, under rocks, for example, are more likely to catch Bd, and periods of stress, perhaps as a result of unusually warm or dry weather, may also trigger outbreaks (and may also increase the tendency of the frogs to aggregate). Researchers discovered that although all amphibians have bacteria and other microbes on their skin that help them to fight off diseases, including fungal infections, the effectiveness of this defence appears to vary from one species to another. Partially immune species can act as reservoirs of infection, or carriers, allowing the fungus to spread throughout a community of frogs. They also found that Bd can survive away from its hosts, at least for limited periods of time, and can be collected from wet rocks on which frogs were previously sitting, for instance.

So far, nobody has definitely established where Bd came from. One theory that is gaining favour among experts is that it has always existed in parts of sub-Saharan Africa, as the fungus was discovered on a museum specimen of an African clawed frog, *Xenopus fraseri,* collected in West Africa as far back as 1933. It was also found in *Xenopus laevis* collected in Uganda in 1934 and in *Xenopus gilli* from the Cape region of South Africa in 1982. Furthermore, the fungus appears not to be fatal to frogs from the southern half of Africa, suggesting that they had acquired immunity to it over a long period of time. Certainly, Bd was present on all clawed frogs collected in the region in 1973 with no apparent ill-effects. Clawed frogs were transported around the world in the middle years of the twentieth century for pregnancy testing and, just as some of the frogs were accidentally introduced to new regions, this too could have been an opportunity for the fungus to spread. Regardless of its origins, Bd has spread to much of the world since the 1980s. The vectors may well be humans, with international travel becoming more popular and eco-tourism, in particular, drawing more people to the kinds of places where frogs live. Scientific researchers and students also move around in such places, some of them monitoring frog populations. Finally, frogs collected for the pet trade may carry the infection to parts of the world where native frogs have no natural immunity.

Urban frogs

Frogs and toads have not adapted to the urban environment to the same extent as some other animals, partly because modern inner city landscapes are, on the whole, dry places. In addition, there are few suitable breeding sites and city noise would probably drown out the sound of their calls. There are a few exceptions however, one of the most common being the Asian bullfrog, *Kaloula pulchra*, which is almost never found outside of villages and towns, where it lives in rubbish dumps, drains and gardens.

Small frogs are often transported accidentally in produce and turn up in parts of the world where they do not belong. The vast majority of these do not survive for long and, since they are usually single animals, they fail to establish themselves even if their new environment suits them. The greenhouse frog, *Eleutherodactylus planirostris*, however, is a West Indian species that has established itself in many of the warmer parts of the United States and Mexico. The coqui frog, *Eleutherodactylus coqui*, another species from the West Indies, was accidentally introduced into Hawaii in 1988 and the males, which make a repetitive, high-pitched call, have become a nuisance to local residents who complain

ABOVE A toad crossing sign near Caerlaverock, Dumfries, Scotland where there is a colony of the rare natterjack toad, *Epidalea calamita* (formerly *Bufo calamita*).

that they cannot listen to the television, or sleep, with windows open. They have largely been eradicated in some places by spraying with citric acid.

In parts of Europe, including the United Kingdom, suburban garden ponds have become one of the most important breeding sites for the common frog, *Rana temporaria*, because so many natural ponds have been filled in and marshy ground drained for agricultural and building development. There are several websites offering advice on how to build frog-friendly garden ponds.

The disadvantages for urban frogs are the dangers that come from pets, inquisitive children, chemical herbicides and pesticides, and traffic. Seasonal breeding migrations are especially hazardous and many frogs and toads die when their migration routes take them across busy roads. In places, tunnels have been built to guide migrating animals safely underneath main roads and in more rural locations 'toad crossing' signs have been erected.

ABOVE The Asian bullfrog, *Kaloula pulchra*, is hardly ever seen outside towns and villages and breeds after heavy rain in drains, ditches and potholes.

Conservation

Conservation of frogs has, in the past, concerned itself with the protection of habitats in which frogs live, mostly in the form of national parks and nature reserves. The recent emergence of Bd as the most serious threat to amphibians around the world has shown that this alone does not work. Measures that have been proposed include precautions to avoid contamination by tourists and researchers, particularly on footwear and equipment, screening frogs that are transported around the world for the pet trade to ensure that infected animals are not moved, and captive breeding. The latter is already underway. In a high-profile project the Panamanian golden frog, *Atelopus zeteki*, has been collected from its natural habitat in the Valle de Anton and moved to breeding facilities in the United States and a further breeding facility will be built in Panama. Another breeding project in Panama involves the endangered Pirri harlequin frog, *Atelopus glyphus*, while in Vietnam several species of frogs are being bred on a large scale in a facility in Hanoi. Other species could be bred in captivity once their requirements are known and stocks of captive-bred frogs could be kept free of Bd by treatment and isolation. This is regarded as a stop-gap measure, however, because there is little point in saving something if it cannot be returned to the wild at a later date. A possible model for reintroduction is that of the Kibansi spray toad, *Nectophrynoides asperginis*, from Tanzania. This small, bright yellow toad became extinct in the wild when the conditions in the gorge in which it lives were affected by the construction of a dam. Individuals were collected from the site while there was still time and bred in captivity by the Wildlife Conservation Society at the Bronx Zoo. The society is working with the Tanzanian authorities to return some of the captive-bred individuals to a nearby habitat, where an artificial sprinkler system will provide conditions similar to those where the toad lived before the dam was built. Additional work is taking place to eliminate chytrid fungus from the gorge.

BELOW Wild populations of the Panamanian golden frog, *Atelopus zeteki*, are infected with chytrid fungus, and specimens have been collected so that they can be bred in captivity and so avoid extinction.

Several other organizations have been set up to sponsor and promote research into frog conservation, which includes programmes to monitor natural populations and working with zoos and aquariums to create an 'amphibian ark', where species that would otherwise become extinct can be maintained and bred. The organization co-ordinating all of these approaches is the Amphibian Specialist Group (ASG). It has initiated projects in many parts of the world and works with the IUCN to identify species that are in need of help. It also publishes *Froglog*, an online magazine that reports on conservation programmes and other aspects of amphibian biology that might be of interest to herpetologists or readers of this book.

10   The Families

AT PRESENT, FROG CLASSIFICATION IS UNDERGOING a major overhaul. In 1984 there were 21 families and this system was relatively stable. Recently, the number of families has increased, not only because new frogs have been discovered but because new techniques are being used to investigate their relationships. This has resulted in the splitting of some large families into two or more (and, less often, the combining of two families into one). A number of schemes have been proposed over the last five years but some degree of stability has been reached at the time of writing, so I have decided to incorporate this recent research, while accepting that further changes are likely. The net outcome is an increase in the number of families to 49, although not all of these are universally recognized.

OPPOSITE The Asian horned frog, *Megophrys nasuta*, from the Megophryidae family (see p. 139).

The scheme I have followed is that of the American Museum of Natural History, which maintains an online database of all amphibian species at the following website: http://research.amnh.org/vz/herpetology/amphibia/index.php. An alternative online resource is maintained by the University of California at AmphibiaWeb: http://amphibiaweb.org/index.html. Note that these two sites use slightly different schemes but both are invaluable resources for anyone interested in amphibian taxonomy, biology or conservation.

Tailed frogs and New Zealand frogs, Leiopelmatidae

This small family contains four New Zealand frogs in the genus *Leiopelma*, and two tailed frogs, *Ascaphus*, from North America, sometimes treated as a separate family, Ascaphidae. *Leiopelma* have traces of tail muscles although the tail itself is absent in adults. They live in isolated parts of New Zealand, two species, *L. archeyi* and *L. hochstetteri*, live on North Island, *L. hamiltoni* lives on Stephens Island in the Cook Straits and *L. pakeka*, on Maud Island in the Marlborough Sounds. The latter species has also been introduced to several other predator-free islands nearby. Fossils of an additional four species have been found, all believed to have become extinct in the last 1,000 years.

They are small to medium-sized, with Hochstetter's frog, at 48 mm (1 ⁹/₁₀ in), being the largest, and have little or no webbing on their feet and no external eardrums. They live in cool forest habitats and have similar breeding habits. *Leiopelma archeyi, L. hamiltoni*

Archey's frog, *Leiopelma archeyi*, from the North Island of New Zealand.

and *L. pakeka* lay terrestrial eggs and the newly hatched tadpoles climb onto the male's back and continue their development there. *Leiopelma hochstetteri* lays its eggs in small puddles of water and they hatch into tadpoles that complete their development without feeding.

The tailed frogs live in the northwestern parts of the United States and adjacent parts of Canada. They were classified as a single species, *Ascaphus truei*, until relatively recently, when the population from the Rocky Mountains was recognized as a separate species, *A. montanus*. The tailed frogs are small to medium-sized, brown or grey in colour, and live along cold, fast-flowing mountain streams. Because of the high oxygen content of the water, most of their respiration is cutaneous and they have small lungs. As their name suggests, the males have a short copulatory organ, the tail, which is an extension of their cloaca, and they use this to fertilize the eggs internally. Mating takes place in the autumn and the female may store the sperm until the following spring, when she lays strings of eggs attached to stones on the stream beds. The tadpoles have a sucker-like disc surrounding their mouth that they use to attach themselves to stones to avoid being swept downstream. Development is very slow and can take up to four years. Subsequent growth is also very protracted and the adult frogs can live for up to 20 years.

Clawed frogs and Surinam toads, Pipidae

A family of 32 species in five genera, four of which, *Hymenochirus*, *Pseudhymenochirus*, *Silurana* and *Xenopus*, live in the southern half of Africa and the other, *Pipa*, in South America. All the members of this family are aquatic, rarely venturing out of the water. They have large, heavily webbed back feet and the *Hymenochirus* and *Pseudhymenochirus* species also have webbing on their front feet. Most species have small black claws on their hind feet and they all have a series of lateral line organs along their flanks. External eardrums are not present and nor are vocal sacs, although they communicate underwater by means of a clicking sound. *Pipa* species have star-shaped, touch-sensitive organs at the tips of their fingers that help them navigate through heavily vegetated water and may also detect prey.

Xenopus laevis was widely used in laboratories until the 1960s for pregnancy testing and other purposes, and has been introduced into several parts of the world such as North, Central and South America, Europe including the United Kingdom, Indonesia and Ascension Island in the Atlantic Ocean. As a result, it has become a pest species in several places, competing with native frogs and adversely affecting fisheries. The Cape clawed toad, *X. gilli*, on the other hand, is one of the rarest frogs; it has a small and shrinking habitat and the remaining populations are in danger of being diluted through hybridization with *X. laevis*.

The *Pipa* species have interesting breeding habits. The male manoeuvres the eggs onto the female's back, where they develop into advanced tadpoles or, in the case of the Surinam toad, *Pipa pipa*, into fully formed toadlets before being released and becoming independent. Other species, notably *Hymenochirus*, lay their eggs at the water's surface after the amplectant pair perform a series of elaborate somersaults. Tadpoles of these species are carnivorous, eating small aquatic invertebrates, whereas those of *Xenopus* and *Silurana* are mid-water filter feeders and typically hang head-down at an angle of about 45° with their tail beating constantly while their mouth gulps water at regular intervals. These tadpoles have long thin barbels at either side of their mouths that are thought to play a part in feeding.

LEFT Dwarf aquatic frog, *Hymenochirus boettgeri*, West Africa.

RIGHT The Mexican burrowing frog, *Rhinophrynus dorsalis*, the only member of its family.

The Mexican burrowing frog, Rhinophrynidae

This family contains only one species, the Mexican burrowing frog, *Rhinophrynus dorsalis*, a bloated, burrowing species with a pointed snout and short legs. It is dark grey or black, with an orange or red line down the centre of its body and blotches of a similar colour on its flanks. It is found in coastal lowlands from extreme southern Texas, south into Central America as far as Costa Rica. As its name suggests, it lives underground for much of its life and has several adaptations for burrowing, such as the pointed snout and spades on its hind limbs. Its main diet is termites and ants, which it catches by rapidly shooting out its tongue. The frog comes to the surface to breed following heavy rains that flood the coastal plains where it lives, and lays its eggs in temporary pools. The tadpoles, which are similar to those in the Pipidae, are filter feeders but may become predatory in their later stages.

Fire-bellied toads, Bombinatoridae

This is a small family containing only eight species, two rarely seen and little-known toads belonging to the genus *Barbourula*, from Southeast Asia, and six better known *Bombina* species, from Europe and Asia. Some authorities have, in the past, included the fire-bellied toads, midwife toads, *Alytes* spp., and the painted frogs, *Discoglossus* spp., in a single family, the Discoglossidae.

The two *Barbourula* species are commonly known as flat-headed toads and are both very rare. *Barbourula busuangensis*, occurs in the Palawan group of islands in the Philippines and the other, *B. kalimantanensis*, in Borneo. They are highly aquatic and adapted to life in streams.

Bombina are much more familiar and are characterized by a warty back and unusual teardrop-shaped pupils. They have colourful undersides and, if threatened, they arch their backs to display the warning colours on their undersides and palms of their feet. One species, the Oriental fire-bellied toad, *B. orientalis*, is widely kept as a pet. Two species with orange undersides, *B. maxima* and *B. microdeladigitora,* are endemic to China and the other three, *B. bombina*, *B. pachypus* and *B. variegata*, are European. Fire-bellied toads typically live in small and shallow bodies of water such as rice paddies in the case of the eastern species and roadside ditches, wheel ruts and even animal hoof-prints in the case of the European species. These habitats are often free from predators. They hang in the water with just the tops of their heads showing and the rest of their bodies floating just below the surface. Fire-bellied toads lay large batches of small eggs which develop in water. The young metamorphose when they are a small size, about the size of house flies.

ABOVE **Yellow-bellied toad,** *Bombina variegata*, Italy.

Midwife toads and painted frogs, Alytidae

Twelve species in two distinct genera form this small family but there are alternative schemes. The name of the family was changed from Discoglossidae to Alytidae because the latter was its original name and so it had priority. Some authorities separate the two genera into separate families, Alytidae and Discoglossidae, while others have, in the past, included them with the fire-bellied toads and placed them in the Bombinatoridae.

The five *Alytes* species are midwife toads and found in Europe and North Africa. They are rather similar and *A. maurus* was previously considered to be a subspecies of *A. obstetricans*. *Alytes muletensis* was only discovered in 1979, on the island of Mallorca in the Balearic Islands, Mediterranean Sea. Male midwife toads carry short strings of eggs wrapped around their hind legs, releasing them into small bodies of water when they are about to hatch. The tadpoles develop to a large size before metamorphosing. All species are small, short-legged toads that have a melodious call that gives them their alternative name of bell toads.

The painted frogs, of which there are seven species, occur in the Mediterranean region. There are four in Spain, Portugal and North Africa, a single species that occurs on Sardinia and Corsica and another endemic to Corsica. The seventh species, *Discoglossus nigriventer*, lived in a small area of Israel but is now presumed extinct due to habitat destruction. Painted frogs are larger than midwife toads, and are more aquatic, have relatively long hind legs making them more agile, and lay large clutches of small eggs in shallow ponds.

RIGHT Mallorcan midwife toad, *Alytes muletensis*.

LEFT Western spadefoot toad, *Pelobates cultripes*, Spain.

Spadefoot toads, Pelobatidae

The Pelobatidae contains a single genus, *Pelobates*, from Europe (*P. cultripes* and *P. fuscus*), the Middle East (*P. syriacus*) and North Africa (*P. varaldii*). The latter species is listed as endangered by the IUCN. They are typically rotund species with smooth skin, large eyes with vertical pupils, and short limbs. The hind legs are equipped with hard, sharp-edged tubercles or spades, which they use for digging backwards into the soil and which give them their common name. They spend much of their lives beneath the surface, thus avoiding high temperatures and dry conditions, emerging onto the surface to feed and breed only when conditions are suitable. They breed in spring, following heavy rain. Amplexus is inguinal and the eggs are laid in thick strings, entangled in aquatic plants. Their tadpoles may reach a very large size, up to 17cm (6 ¾ in) in some cases. Some authorities consider the American genera, *Scaphiopus* and *Spea*, to be part of the same family but here they are placed in a separate family, the Scaphiopodidae.

Horned frogs, litter frogs and their relatives, Megophryidae

The Megophryidae contains ten genera and 149 species of Asian frogs (sometimes referred to as toads). They used to be included within the spadefoot family, Pelobatidae, and many species have a spade-like projection on their hind limbs. The family is distributed from southern China to Southeast Asia, including the Philippines.

The members of this family are so diverse that it is difficult to generalize about them. Several species are remarkable for appearing like dead leaves, with projections over their

eyes and on their snouts, leaf-like shape and coloration. Most notable among these, and the best known, is the Asian horned frog, *Megophrys nasuta* (p.132) The genus has four other species, but was formerly larger; many former species have been moved to different genera, the largest of which is *Xenophrys* with 37 species. Five *Brachytarsophrys* species are rather similar although not all have the eyebrow extensions. *Borneophrys edwardinae* is the sole member of its genus as it was recently removed from *Megophrys*. These species, grouped together into the subfamily Megophryinae, are relatively large and have large mouths. The other species, placed in the subfamily Leptobrachiinae, are small to medium-sized frogs, known collectively as litter frogs. *Leptobrachium* (21 species, known as large-eyed litter frogs) and *Leptobrachella* (7 species of dwarf litter frogs) are smaller, while the 22 *Leptolalax* species are known as slender litter frogs. In addition, 17 species of *Oreolalax* and 18 species of *Scutiger*, are predominantly found in China, and are similar to each other.

All the species in this family, as far as is known, lay their eggs in streams. *Megophrys* and *Xenophrys* species lay their eggs in the pools and backwaters that form along shallow streams and their tadpoles have modified, funnel-shaped mouths for feeding from the water's surface, while those that live in fast-flowing streams have sucker-like mouths for clinging on to rocks. Tadpoles of smaller species, such as *Leptobrachium* and *Leptobrachella*, have elongated bodies so that they can live in the spaces between rocks and shingle on stream beds.

Parsley frogs, Pelodytidae

There are just three species in this family, two from southwestern Europe, *Pelodytes ibericus* and *P. punctatus*, and one from the Caucasus Mountains, *P. caucasicus*. There is a gap of several thousand kilometres between the range of the latter species and the former two species. They are all small, nocturnal, terrestrial frogs that live in woodlands or fields but move to small ponds, ditches and marshes in sunny situations to spawn. Their common name comes from scattered small green markings on the dorsal surface of the two western

LEFT Parsley frog, *Pelodytes punctatus*, Spain.

species. They start to breed in spring and may continue throughout the summer. Amplexus is inguinal and the female winds strings of up to about 350 eggs (less in the case of *P. caucasicus*) around the stem of an aquatic plant as they are laid. As she lays several times in a season, the total number of eggs can be over 1,000. The tadpoles develop in water and the newly metamorphosed frogs can reach breeding size within a year.

American spadefoot toads, Scaphiopodidae

The American spadefoot toads include three species in the genus *Scaphiopus*, *S. couchii*, *S. holbrookii* and *S. hurterii*, and four in the genus *Spea*, *S. bombifrons*, *S. hammondii*, *S. intermontana* and *S. multplicata*. They occur from southern Canada to the highlands of central Mexico. All American spadefoot toads are similar in general appearance to their European counterparts with stout bodies, wide heads, short limbs and hardened spades on their hind feet. Some species live in deserts, where they spend much of the year below ground, emerging to feed during damp weather, and forming explosive breeding groups after heavy rains. In less arid places, however, they are more often seen on the surface and have a seasonal breeding cycle. Species from the most arid regions have an accelerated life-cycle, breeding in temporary pools within days, or hours, of summer rains. Their eggs hatch in a matter of hours and the tadpoles develop rapidly, as quickly as 6–8 days in the case of Couch's spadefoot toad, but more commonly in 24–32 days.

BELOW Couch's spadefoot toad, *Scaphiopus couchii*, Arizona.

The Table Mountain ghost frog, *Heleophryne rosei*, is confined to the eastern slopes of Table Mountain, South Africa, where its habitat is being degraded by the planting of exotic trees.

Ghost frogs, Heleophrynidae

There are only six species in this family, five belonging to the genus *Heleophryne* and one recently moved from this genus to a new one, *Hadromophryne*. They are restricted to South Africa. The five *Heleophryne* species live in the streams that cascade down the rocky hillsides and gorges in the Cape region or along the south coast, while the Natal ghost frog, *Hadromophryne natalensis*, ranges further east, into northeastern South Africa.

They are medium-sized frogs with large eyes and have toes that end in expanded, truncate tips which they use to cling to the wet rock faces in waterfalls and torrents, where they live. They lay their eggs in batches of about 50–200, with only a slight variation between the species, sticking them to the undersides of rocks in backwaters or scattered among damp rocks and moss at the edges of streams and seeps. The tadpoles have large flattened heads and a large oral disc that they use to cling to rocks and prevent themselves from being swept away.

Two species are listed as critically endangered. Hewitt's ghost frog, *Heleophryne hewitti*, lives in an area that has recently been planted with exotic pine trees, which restrict the flow of water in the streams that run through the area. In addition, exotic fish have been introduced into nearby streams in the area. The Table Mountain ghost frog, *H. rosei*, lives only on the eastern flanks of Table Mountain, Cape Town, which have also been planted with conifers, causing their streams to dry up. The Cape Region and the south coast has seen rapid development in recent years, with the building of residential and holiday homes as well as greatly increased tourism and this is also having a negative impact on the frogs' habitat.

Seychelles frogs, Sooglossidae

There are only four species in this family and they are divided between two genera, *Sechellophryne* and *Sooglossus*, with two species each, and are endemic to the Seychelles group of islands in the Indian Ocean. These are very small frogs that live among leaf-litter on forest floors. *Sooglossus thomasetti* occurs in the vicinity of forest streams. All species lay their eggs on the forest floor. Those of *Sechellophryne gardineri*, undergo direct development and the females remain with the eggs until they hatch. The size of newly hatched young has been likened to grains of rice. The eggs of *Sooglossus sechellensis* hatch into non-feeding tadpoles which climb onto the back of the female to complete their development. Details of the natural history of the other two species, one of which *Sechellophryne pipilodryas* has quite recently been discovered, is unknown. A third genus, *Nasikabatrachus*, from India, was included in the family but has now been moved to a family of its own, the Nasikabatrachidae.

The Indian purple frog, Nasikabatrachidae

This family contains a single species, the remarkable Indian purple frog, *Nasikabatrachus sahyadrensis*, from the Western Ghats of southern India. It is purplish-brown in colour, about 7 cm (2 ⅘ in) long and has a pointed snout, bloated body and short limbs. It spends most of its time below the ground in montane forests, feeding on termites and emerging only in the monsoon to breed in temporary pools. This bizarre frog was first described in 2003 and placed in a new family. It was later found to be closely related to the Seychelles frog and was moved to the Sooglossidae for a time but its distinctiveness and geographical separation have resulted in a move back to the Nasikabatrachidae.

LEFT The strange Indian purple frog, *Nasikabatrachus sahyadrensis*, discovered in 2003 and placed in a family of its own.

Chilean toads, Calyptocephalellidae

This small family, formerly part of the Leptodactylidae, has only two genera and four species, all from southern Chile. The helmeted water toad, *Calyptocephalella gayi*, is restricted to wet habitats along the Pacific coast of Chile, where its numbers are declining. It is a large toad with a wide gape, and is almost entirely aquatic. It was previously known as *Caudiverbera caudiverbera*.

The three *Telmatobufo* species occur in southern Chile where they live in mountain streams. Their tadpoles are highly specialized for living in torrents and waterfalls, with flattened bodies and very large oral discs that help them grip to the substrate. The adults are stocky, with wide heads and large warts on their backs and sides. Otherwise, their natural history is unknown. All three species are rare and known only from a small number of specimens. The IUCN lists *T. australis* as 'vulnerable', *T. bullocki* as 'critically endangered' and *T. venustus* as 'endangered'. The latter species was described in 1899 and rediscovered 100 years later.

Australian ground frogs, Limnodynastidae

BELOW Moaning frog,
Heleioporus eyrie, Walpole,
Western Australia.

A family of 44 species in eight genera, the Limnodynastidae are restricted to Australia and New Guinea. They are terrestrial species that live in a variety of habitats from swamps to deserts. The genus *Limnodynastes*, with 11 species, are variously known as marsh frogs or banjo frogs, named for their habitat and their call. Collectively, they are found throughout Australia. *Lechriodus*, with one species in Australia and three in New Guinea, are similar and members of both species lay their eggs in foam nests floating on shallow water. Ten species of *Neobatrachus* and the two rather similar *Platyplectrum* species are stout, burrowing species from dry areas, and several species spend extended periods below ground to avoid drought. Similarly, the spadefoot toads, *Notaden*, are globular-bodied burrowers from deserts that are equipped with calcified spades on their hind feet that help in burrowing. The holy cross toad, *N. bennetti*, is a particularly colourful species with bright yellow patches on its back. Yet more burrowing species are placed in the genus *Heleioporus*, of which there are six, but these frogs are not found in deserts. They live in burrows near watercourses and lay their eggs in foam nests inside the burrows following heavy rain in spring and summer. Five occur in extreme southwestern Australia and the other one is from the southeast of the country. Six species of *Philoria* have very small ranges in damp grassland and woodland in the extreme east of the country, and all are classed as endangered or critically endangered. The remaining species is the tusked frog, *Adelotus brevis*, in which males have long tooth-like tusks on the lower jaw that they use to bite other males when they fight each other at breeding time.

Australian froglets and toadlets, Myobatrachidae

The Myobatrachidae consists of 84 species in 13 genera. They are exclusively Australasian. *Mixophyes hihihorlo* is endemic to New Guinea and two others, *Crinia remota* and *Uperoleia lithomoda*, are found in both New Guinea and Australia. The other 81 species are endemic to Australia. They range from tiny 20 mm (¾ in) long, moss-dwelling *Crinia*, *Geocrinia* and *Pseudophryne* species, sometimes known as froglets or toadlets depending on the species, to the large barred frogs, *Mixophyes* species, which can grow to over 100 mm (4 in) and will eat other frogs. Similarly, their shapes vary from slender, long-legged species like the torrent frogs, *Taudactylus*, to robust, short-legged ones such as the turtle frog, *Myobatrachus gouldii*, and their colour varies from the brightly coloured corroboree toadlets, *Pseudophryne corroboree* and *P. pengilleyi*, to the duller coloration of the majority of the species, which rely on camouflage. In short, this is a varied family about which it is difficult to generalize.

The largest genus, *Uperoleia*, contains 26 small species, mostly from grassland and lightly wooded areas, although some live in deserts. A number have very limited ranges in the arid Kimberley region of northwestern Australia. The *Crinia*, *Geocrinia* species and the single species of *Paracrinia* live in damp places, such as flooded woodlands and peat swamps. Several species are very similar to each other and, in addition, they are

BELOW Northern barred frog, *Mixophyes schevelli*, Atherton Tablelands, Queensland, Australia.

often highly polymorphic and so identification can be difficult. They live at the fringes of pools, streams and seeps and lay their eggs under stones, among leaves or in soil that later becomes flooded. The toadlets, *Pseudophryne* species, are small terrestrial frogs, some of them colourful. Most species lay their eggs in damp places and the eggs hatch when the area becomes inundated. Males often remain with the eggs. Torrent frogs, *Taudactylus*, live in fast-flowing streams within the coastal rainforests of Queensland, and the tadpoles of some (perhaps all) species have disc-like suckers on the undersides of their head so that they can cling to rocks. Four out of the six species are endangered and the Mount Glorious torrent frog, *T. diurnus*, is probably extinct. Two gastric-brooding frogs, *Rheobatrachus silus* and *R. vitellinus*, are from the same region. Discovered as recently as 1973 and 1984, they are almost certainly already extinct. Females of these species swallowed their eggs or tadpoles and development continued in their stomachs.

Remaining species in the family are placed in genera with only one or two species. Two sandhill frogs, *Arenophryne rotunda* and *A. xiphorhyncha*, and the turtle frog, *Myobatrachus gouldii*, are ant- and termite-eating, burrowing species from dry sandy places that live and breed underground. The marsupial frog, *Assa darlingtoni*, is a small species in which the male has pouches in his hips into which the tadpoles wriggle and complete their development there. Nicholl's toadlet, *Metacrinia nichollsi*, is a small, rare species from Western Australia that lives in karri forests and whose eggs develop directly into small adults, and the sunset frog, *Spicospina flammocaerulea*, is a colourful species with a bright yellow underside that lives in a few isolated peat swamps and lays its eggs in shallow water among mosses and tussock grass.

Marsupial frogs, Hemiphractidae

As currently understood, this family consists of 93 species in five genera. There are several other arrangements and the relationships of the genera have not been fully resolved, so future changes are possible. All species in this family were previously included in the Hylidae.

Six *Hemiphractus* species are strange-looking frogs with bony heads and pointed projections over their eyes and snout. They live on the forest floor, where they are difficult to see among dead leaves, but may climb into low shrubs at night. They have large mouths and gape widely if threatened, displaying the bright yellow interior. Very few specimens have been collected despite a large composite range from Panama to Brazil. In common with the members of two other genera – *Stefania*, with 18 species, and *Cryptobatrachus* with six – the females carry their eggs on their backs and they hatch directly into froglets. Two other genera, the marsupial frogs, *Gastrotheca* with 58 species and *Flectonotus* with five, also carry their eggs but they are fully or partially enclosed in pouches. In *Flectonotus* and some of the *Gastrotheca* species, the eggs hatch as advanced tadpoles which the female deposits in water, often using the long toes of her hind feet to scoop them out of the pouch. Other *Gastrotheca* species, however, retain the developing tadpoles in their pouch until they have metamorphosed.

Gastrotheca and *Flectonotus* occur from Panama to northern Argentina and are sometimes placed in a separate family, the Amphignathodontidae, while *Cryptobatrachus* and *Stefania* are restricted to northern South America and are also sometimes placed in a separate family, the Cryptobatrachidae.

Tepui frogs, Ceuthomantidae

This is a new family of three species, all in the genus *Ceuthomantis*, and are found in the flat-topped mountains, or tepuis, in northern South America. They are closely related to the Brachycephalidae, Craugastoridae, Eleutherodactylidae and Strabomantidae and lay directly developing eggs. No other aspects of their natural history are known.

Rain frogs and toadlets, Brachycephalidae

Forty-four species in two genera, *Brachycephalus* and *Ischnocnema*, currently make up the Brachycephalidae, although this family has been subject to much change in recent years. This family is found in Brazil and northern Argentina and its members are small, forest species. Seven of the 12 species of *Brachycephalus* have been described from Brazil since 2000. The species in this genus are so small that they have a reduced number of digits, with two or three on their front feet and three on their hind feet. *Brachycephalus pitanga* is among the smallest species in the family, and one of the smallest of all frogs, at 10–14 mm (²/₅–³/₅ in), although *B. didactylus* and *B. ephippium* may be equally small. All are bright orange frogs with toxic skin secretions. The 32 members of the other genus, *Ischnocnema*, are also very small, brown in colour and live among leaf-litter. They are only found in Brazil, where many of them are threatened by the destruction of the Atlantic coast forests. All members of the family lay terrestrial eggs that develop directly into small froglets.

Rain frogs, Craugastoridae

The Craugastoridae contains 114 species in two genera, *Craugastor* with 112 species and *Haddadus* with two. Their relationships are unclear and they have, at various times, been included in the Brachycephalidae, Eleutherodactylidae and Leptodactylidae. They have a combined range from southern Arizona and central Texas, through Central America and as far south as eastern Brazil and northern Ecuador in South America. These are small frogs that live among leaf-litter, rocky slopes and in caves. They are mostly brown in colour and all species have direct development. *Craugastor augusti* is found in the United States, where it is known as the barking frog on account of its yelping call.

ABOVE The big-headed rain frog, *Craugastor megacephalus*, from Central America.

Rain and litter frogs, Eleutherodactylidae

The Eleutherodactylidae contains 201 species in four genera, one of which, *Eleutherodactylus* contains 185 species. There has been much re-arrangement among the species in this family, the members of which were formerly placed in the Brachycephalidae and, before this, in the Leptodactylidae. In a further complication, some of former members of the genus *Eleutherodactylus* have been moved to other families such as Strabomantidae and Craugastoridae.

Its members occur from the southern United States to central South America, including many Caribbean islands. *Eleutherodactylus* and related species are small to medium-sized frogs, mostly brown but sometimes attractively coloured, that live on the ground or in trees. They are sometimes known as rain frogs or, in some places, coquis, after the call they make. All species have direct development. *Eleutherodactylus jasperi* (now probably extinct) from Puerto Rico, has internal fertilization and retains its eggs in its oviduct, giving birth to live young. Another Puerto Rican species, *E. coqui*, has internal fertilization but lays eggs. Direct development, coupled with their sheer numbers, has facilitated their accidental introduction to many places, as eggs are moved inadvertently in soil, moss or among plant roots. Adults are often transported around the world in produce such as fruit and may also become established in suitable places. The Cuban *E. planirostris*, for example, is now found in Florida and several other states in the American southeast, where it is known as the greenhouse frog, as well as in Jamaica, Guam and Hawaii, where it is considered a pest. Many make a plaintive peeping call. Nine species in *Diasporus* have recently moved from *Eleutherodactylus*. The other genera are *Adelophryne* with six species from the Guiana Shield and *Phyzelaphryne* with a single species from southern Brazil. Many more of these small, inconspicuous and rather nondescript frogs must surely await discovery.

ABOVE An *Eleutherodactylus* spp. introduced to the UK accidentally in bananas. Small West Indian species such as this often turn up in fruit but do not normally survive for very long.

Rain frogs, Strabomantidae

The relationships within this large family are poorly known. The 562 species in 17 genera have been included in various other families in the past and many new species have been described recently. They have a large combined range from eastern Honduras, through Central America and into South America, where they occur down the Andes and throughout the Amazon Basin, and into the Atlantic forests of Brazil. The largest genus by far is *Pristimantis*, with 434 species. It has a range similar to that of the family and its members are variable, with cryptic, brown, forest-floor species and bright green arboreal species. *Pristimantis gaigei* has two bright orange stripes down its back and is thought to be a mimic of the poisonous *Phyllobates* species that share its habitat. Another species, *P. altae*, is almost entirely black in colour except for a bright red or orange spot in its groin, but the purpose of this coloration is not known. Sixteen other genera have from one to 22 species. They are small, forest-dwelling frogs, some of which are arboreal. As far as is known, all species lay terrestrial eggs and have direct development.

Tree frogs, Hylidae

The tree frogs, Hylidae, form a large family of 891 species, with an extensive geographical range. They occur throughout North and South America, southern Europe, North Africa, eastern Asia including Japan, southern China and neighbouring countries, Australia and New Guinea. Their numbers and large distribution is reflected in their diversity. Although they are predominantly tree frogs, with slender bodies and adhesive toe pads, some members of the family are also found in a range of other habitats.

They are divided into three well-defined subfamilies. The largest is Hylinae, containing 636 species, and covering most of the range of the family as a whole, except the Australian region. By far the largest number come from Central and South America, where many of the species formerly placed in the genus *Hyla* have been moved to a number of smaller genera, including several new ones, so many of the names are unfamiliar. They include *Scinax*, with 98 species (some of them previously known as *Ololygon* species), and *Hypsiboas*, a genus that contains 82 species, including the gladiator frogs, males of which build mud basins where they breed and guard their eggs. Forty-two species of *Plectrohyla* are known as spike-thumb tree frogs because of a spine on their hands which they use in combat with other males. This characteristic is shared by the 11 members of the genus *Ecnomiohyla*, known as marvellous frogs because of their spectacular markings. They also have heavily webbed front and hind feet. Seven species of casque-headed tree frogs, *Trachycephalus*, secrete a poisonous, milky substance from the skin and are sometimes known as milk frogs.

The type genus *Hyla* contains 34 species, distributed across central and southern Europe, North Africa, eastern Asia and North America. These are typical tree frogs, including four European green tree frogs, all very similar to each other, the Japanese tree frog, *Hyla japonica*, and many of the familiar North American species such as the green tree frog, *Hyla cinerea*, the grey tree frogs, *H. chrysoscelis* and *H. versicolor*, and the barking tree frog, *H. gratiosa*.

ABOVE LEFT Barking tree frog, *Hyla gratiosa*, Florida.

ABOVE RIGHT Pacific tree frog or Pacific chorus frog, *Pseudacris regilla*, California.

Apart from these arboreal tree frogs there are a number of more terrestrial species, including the cricket and chorus frogs, *Acris* (three species) and *Pseudacris* (18 species). The toe pads of these frogs are greatly reduced in size. Some members of the wide-ranging genus *Smilisca*, with eight species, burrow to avoid dry weather, as does the single species of *Triprion*, the casque-headed tree frog, *T. petasatus*, which comes from Central America.

Breeding in most species takes place in water, with clusters of eggs attached to aquatic plants but a small number breed in bromeliads and tree holes. Some species of the bony-headed tree frog genus, *Osteocephalus*, such as *O. deridens* and *O. oophagus*, lay their eggs in bromeliads and the female returns to the site at regular intervals to lay infertile eggs for them to eat. Other species, such as *O. taurinus*, spawn in shallow temporary pools and their tadpoles feed on eggs of their own and other species. The crowned tree frog, *Anotheca spinosa*, spawns in tree holes, attaching its eggs to the inside of the cavity, just above the water's surface. When the tadpoles hatch they slide down into the water and the mother returns to the site to lay infertile eggs for them to eat.

The 11 species in the genus *Pseudis*, from Central and South America, are known as paradoxical frogs because their huge tadpoles can grow up to four times the length of the adults into which they metamorphose. They are all highly aquatic species, with heavily webbed hind feet, not superficially like tree frogs at all, and they were previously placed in a separate family, the Pseudidae. *Pseudis paradoxus* aestivates in hardened mud in parts of the Chaco region of Argentina, Bolivia and Paraguay, where ponds and lakes frequently dry up.

The Phyllomedusinae contains 59 species in five genera. They have vertical pupils, distinguishing them from all the other American members of the Hylidae, which have horizontal pupils. They are large tree frogs, highly arboreal and mostly green in colour, sometimes with brightly coloured flash markings on their groin and flanks. With one or two exceptions, they lay their eggs on leaves overhanging forest pools and some species fold the margins of the leaves over their spawn to form a tube. The largest genus is *Phyllomedusa*, with 31 species, often known as leaf frogs or monkey frogs. They rarely leap but progress with a slow, deliberate gait, gripping thin branches and stems with opposed digits. *Agalychnis* contains five species of gliding frogs, with heavily webbed front and hind feet. Two species, *A. callidryas* and *A. natator*, have bright red eyes and the former is an iconic frog associated with the Costa Rican rainforest (although its range extends into neighbouring countries). *Agalychnis moreletii* is classed as critically endangered and *A. annae* as endangered. Five *Phrynomedusa* species come from Brazil. *Phrynomedusa marginata* is unusual in being brown rather than green and hiding its eggs in crevices. *Phrynomedusa fimbriata* is probably extinct. The remaining species are seven *Phasmahyla*, all from Brazil, and two spectacular frogs in the genus *Cruziohyla*, *C. calcarifer* and *C. craspedopus*, which are large, brightly coloured Central American species with heavily webbed feet that live high in the forest canopy and are rarely seen.

BELOW Small red-eyed leaf frog, *Agalychnis saltator*, Costa Rica.

The members of the subfamily Pelodryadinae are endemic to Australia, New Guinea and neighbouring Indonesian islands. Three species, *L. aurea*, *L. ewingii* and *L. raniformis*, have been introduced to New Zealand. All 196 species are sometimes placed in the genus *Litoria*, although some authorities prefer to recognize 13 species of water-holding frogs as belonging to a separate genus, *Cyclorana*, and 25 species of lace-lid tree frogs to another genus, *Nyctimystes*. In contrast to the other members of the subfamily, the lace-lids have vertical pupils and a heavily patterned lower eyelid, which gives them their common name. Although most species are typical, slender-bodied tree frogs, the green tree frog, *L. caerulea* and the magnificent tree frog, *L. splendida*, are large, bulky species that cannot support themselves on vertical surfaces when they are fully grown. Both of these, and another large species, the white-lipped tree frog, *L. infrafrenata*, are associated with human dwellings and are often found around houses, particularly in toilets. *Litoria nasuta* is a long-legged terrestrial species known as the rocket frog on account of its ability to make great leaps of over 1 m (40 in). Thirteen terrestrial species sometimes placed in the genus *Cyclorana* burrow into the substrate to avoid dry weather if necessary. Breeding among most of the species in this subfamily takes place in ponds but some species, such as the endangered waterfall frog, *L. nannotis*, common mist frog, *L. rheocola*, a number of species related to *L. pratti*, from New Guinea, and the lace-lid tree frogs, are specialized stream breeders and their tadpoles have flattened bodies and large discs on their undersides for clinging to rocks. A small number of species from New Guinea, including *L. iris* and some of its close relatives, lay their eggs on leaves overhanging ponds, drainage ditches and streams and the tadpoles drop into the water when they hatch.

The Tukeit Hill frog, Allophrynidae

The single species in this genus, *Allophryne ruthveni*, is something of a mystery. It is often placed in the Centrolenidae, or glass frogs, to which it appears to be closely related, but has also been included in the Hylidae and Bufonidae in the past. Although it has toe pads, the underlying structure of these is different from those of members of the Hylidae; they more closely resemble those of the glass frogs but there are other differences that make its relationships far from clear and the most conservative approach seems to be to treat it separately. It occurs in northern South America, extending down into Brazil. It grows to 25–30 mm (1–1 $\frac{1}{5}$ in) and its back is covered with pointed tubercles. It is terrestrial or arboreal and has been found in the central water-filled centres of terrestrial bromeliad plants. It breeds in forest pools and its eggs and tadpoles develop in water.

Glass frogs, Centrolenidae

There are 145 species and 11 genera in this family. The family has been revised recently and several new genera have been named. It is divided into two subfamilies, the Centroleninae, with nine genera, six of them newly described, and the Hyalinobatrachinae, with two genera, one recently described.

The Centrolenidae are commonly called glass frogs because the skin on their underside is transparent, allowing their internal organs to be seen. They have expanded toe discs, large eyes with horizontal pupils and most species are green, with small spots of white or black. They are arboreal inhabitants of rainforests in Central and South America, and breeding takes place on leaves above forest streams and rivers. Males call from suitable vantage points, which might be several metres above the ground, and the female lays her eggs on the upper or lower surface of a leaf. The males of many species stay with their eggs and may attract additional females, so they might be guarding two or three clutches, which contain two to 30 eggs, at the same time. When they hatch, the tadpoles drip down into the water below to continue their development. The giant glass frog, *Centrolene buckleyi*, may breed in terrestrial bromeliad plants though this species, which lives above the tree-line in Ecuador, is almost extinct. Another large species, *C. geckoideum*, lives in the spray zones behind and alongside waterfalls and there are reports that it attaches its eggs to rock faces.

BELOW Glass frogs are well named. The internal organs of this northern glass frog, *Hyalinobatrachium fleischmanni*, are clearly visible through its underside.

Tropical American grass frogs, Leptodactylidae

This family has been the subject of recent revisions, some of which have not been completely resolved. It was formerly much larger but was recently divided into several separate families, leaving 99 species in four genera with a combined distribution of the whole of South America, southern Central America extending north along the Gulf of Mexico as far as southern Texas.

Leptodactylus, which has 88 species, is the largest genus. Its members are superficially similar to the ranids from the northern hemisphere which, in ecological terms, they replace. These frogs can be found almost anywhere within the family's range, living in a variety of damp habitats, often close to water. They lay their eggs in foam nests, which may float on the surface of ponds or in burrows or depressions nearby. *Leptodactylus pentadactylus* is the Smokey Jungle frog, with a large range from Central America to Amazonian Brazil. It may grow to 180 mm (7 in) in length and is eaten by humans in parts of its range. Three other small genera, *Hydrolaetare* with three species, *Paratelmatobius*, with seven species, and *Scythrophrys*, with one species, complete the family.

Horned frogs and water frogs, Ceratophryidae

The Ceratophryidae contains 86 species in six genera and the family is divided into three subfamilies. It was formerly part of the Leptodactylidae. Members of this South American family are mostly terrestrial, but some are aquatic. They all have aquatic eggs and tadpoles, some of which are predatory.

The Batrachylinae consists of two genera and 14 species from Chile and Argentina. Patagonian frogs in the genus *Atelognathus*, live on Andean slopes and lakes in Chilean and Argentinean Patagonia. They are poorly known, and some are endangered. *Atelognathus patagonicus* lives under stones at the edge of lakes and appears to breathe through its highly vascular skin so in this respect it is similar to some of the *Telmatobius* species. Four *Batrachyla* species have a similar distribution. At least some species lay their eggs in damp moss or vegetation near water.

The Ceratophryinae includes 12 species in three genera. *Ceratophrys* are large, wide-mouthed species known as horned frogs. They are well camouflaged and ambush their prey, which often include other frogs as well as small mammals, reptiles and large invertebrates, from a concealed position. They are popular pets and are known in the pet trade as pac-man frogs. The other two genera, *Chacophrys* and *Lepidobatrachus*, are similar in many respects, including their feeding habits but whereas *Ceratophrys* and *Lepidobatrachus* tadpoles are carnivorous, those of *Chacophrys* are typical grazers, eating algae and other plant material. All 12 species have hardened spades on their hind feet for digging, and *Chacophrys* and *Lepidobatrachus* live in arid regions where they sometimes need to burrow down into the mud to aestivate.

A single genus, *Telmatobius*, with 60 species, forms the Telmatobiinae. They live in the Andes, from Ecuador to Chile, many in lakes at high altitudes, where they are totally aquatic. The Lake Titicaca frog, *T. culeus*, is remarkable for its loose, flabby skin that allows it to extract most of the oxygen it needs from the water.

LEFT Horned frog, *Ceratophrys cornuta*, South America.

Cycloramphidae

This newly created family has no common name. Most of the species were previously included in the Leptodactylidae but the mouth-brooding frogs, *Rhinoderma* species, which are now included in this family, were previously given a family of their own, the Rhinodermatidae. The Cycloramphidae is divided into two subfamilies, the Alsodinae and the Cycloramphinae, with nine and four genera respectively, and an additional species, *Rupirana cardosoi*, whose place is uncertain. In total there are 101 species in the family, all South American, occurring from northwestern Brazil, through Bolivia and into Chile and Argentina. Most are small and dull in colour. The natural history of many species is sketchy but at least some, such as *Eupsophus calcaratus*, lay their eggs on damp ground beneath logs and rocks. They hatch into tadpoles and swim away after the ground becomes inundated after rain. There is some evidence that one of the parents remains with the eggs until they hatch. *Zachaenus* species lay small clutches of eggs on land and the tadpoles develop in the jelly mass, feeding entirely on their yolk, whereas the *Thoropa* species, as far as is known, lay their eggs on damp rock ledges and the elongated tadpoles live in cascades and torrents.

The most interesting species are the mouth-brooding frogs, *Rhinoderma*, from southern Chile and Argentina, which are remarkable for their mouth-brooding habits (see p.102). Darwin's frog, *Rhinoderma darwinii*, is small, from 2.5 to 3.5 cm (1 in to 1 ⅜ in) in length, quite rotund, and has small triangular flaps of skin on its heels and another, more prominent one, on the tip of its snout. It may be brown or green in colour, with a black and white underside. When alarmed it will leap into the water and pretend to be dead, floating upside down. There is one other species in the genus, *R. rufum*, also from Chile but it has not been seen since 1980 and may be extinct. The ranges of the two species overlap, or overlapped, but *R. rufum* differs in having more heavily webbed toes. Although male *R. rufum* apparently place their tadpoles in their vocal pouch, they release them before they have metamorphosed.

RIGHT Chiloe Island ground frog, *Eupsophus calcaratus*, Puerto Vallarta, Chile.

LEFT **The Tungara frog,**
Engystomops pustulosus, seen
here calling in shallow water,
is a common member of the
Leiuperidae.

South American foam-nesting frogs, Leiuperidae

Formerly included within the Leptodactylidae, the Leiuperidae includes 79 species in seven genera from South and Central America as far north as central Mexico. They are small to medium-sized terrestrial frogs, most of which produce foam nests on the surface of shallow water, often after heavy rain. Most are grey or brown frogs with rough skin although *Eupemphix nattereri* and some *Pleurodema* species have prominent eye-spots on their rumps that they use to intimidate predators. Members of *Engystomops* genus are known as foam frogs and *E. pustulosus* is a common and frequently heard species from Mexico to northern South America, where it congregates in large numbers after rain. Its calls carry for long distances. Eleven species of *Pseudopaludicola* differ from other members of the family by laying eggs directly in the water, without foam nests.

Toads, Bufonidae

The toads are a large family of 550 species divided into 48 genera. The family's range includes every continent except Antarctica. It does not naturally occur in Australia but one species, the cane toad, was introduced there and its range is spreading. The family has been the subject of much recent revision, particularly the former genus *Bufo*, which has been reduced from over 250 species to just 17. *Bufo* species, as now understood, occur in Europe, North Africa, temperate Asia and parts of Southeast Asia. Some of the more recently described, or resurrected, genera that include species formerly in *Bufo* are

TOP LEFT The South American rococo toad, *Rhinella schneideri*, grows to 250 mm (10 in) long making it one of the largest species of toad.

TOP RIGHT The purpose of the strange protuberance on the snout of this Peruvian toad is unknown and it may simply serve to break up its outline. The identity of the toad is uncertain although it is thought to be *Rhinella dapsilis*.

ABOVE Southern toad, *Anaxyrus terrestris*, Florida.

Rhinella, with 86 species distributed from southern United States to South America. They include the marine, or cane toad, *R. marina*, which has additionally been introduced to many other regions, notably Hawaii, the Philippines, New Guinea and Australia. The genus contains several other large species, such as *Rhinella schneideri* (*Bufo paracnemis*) as well as smaller South American species such as the long-snouted *R. dapsilis* (*Bufo dapsilis*). Other new genera include *Incilius* (34 species from North, Central and South America), *Anaxyrus* (22 species from North America), *Amietophrynus* (39 species from Africa) and *Duttaphrynus* (29 species from East Africa and Asia). All these species are nocturnal, terrestrial species with warty skin that often have large parotid glands behind their eyes. Most species are brownish in colour but some are olive or even reddish and their markings consist of darker spots or blotches. Many species have a pale yellow or white line down the centre of their back. These former *Bufo* species lay their eggs in single or double strings in ponds and other small bodies of water and their tadpoles develop quickly and metamorphose in three to ten weeks. Desert species sometimes use temporary pools and the European natterjack spawns in shallow dune slacks in the north of its range.

There are eighty-nine species in the *Atelopus* genus from Central and South America. Nearly every species of *Atelopus* is endangered, critically endangered or extinct, with at least three species becoming extinct since the 1980s. They are sometimes known collectively as harlequin toads, although despite this name not all are colourful, some Andean species being black; their alternative name is stubfoot toads. They have smooth skins and lack a visible tympanum. *Atelopus varius* from Central America, is a wonderfully colourful species that can have a whole range of bright colours, including yellow, orange and red, in the form of spots or stripes, on a jet black background. The Panamanian golden frog , *A. zeteki*, is bright orange with small black markings. It lives along mountain streams in Panama. Both these species are critically endangered as a result of infection by chytrid fungus. Twenty-five species of the genus *Melanophryniscus* occur in the southern half of South America, from Argentina to southern Bolivia and Brazil, Paraguay and Uruguay. They are unlike other toads, being black or dark brown in colour with bright red patches of skin on their undersides and on the palms of their hands and feet, which they display when they are disturbed by arching their back and raising their limbs, a form of 'Unkenreflex' similar to the fire-bellied toads, *Bombina*. Some species also have red or yellow spots on their backs and are known as bumblebee toads. They are small, about 25–30 mm (1–1 ⅛ in), and live near mountain streams. Males call in the morning and the egg-masses are attached to the stems of aquatic plants.

Nine species of *Oreophrynella* are endemic to the dramatic sandstone tepuis of the region bordering Guyana and Venezuela. Species occur on the flat-topped summits, along with many other interesting and unique plants and animals. They are small, black and diurnal, and known from very small areas. The Venezuelan tepui toad, *O. nigra*, is the best-known species and was featured in a sequence on a television series called *Life*, rolling and bouncing down a mountainside to escape a predatory spider. The other eight species of the genus were all described fairly recently. The Mount Roraima tepui toad is endemic to the mountain of the same name and *O. seegoini* is endemic to Maringma tepui. The life-cycles of these species are unknown but they are thought to lay terrestrial eggs that develop directly.

Other noteworthy genera include the Asian slender toads, *Ansonia*, of which 25 species inhabit Asia. Although unmistakably bufonids, with warty skins and large parotid glands, they are of a more slender build and have long, thin legs and digits. They live alongside streams in primary rainforests and their elongated, flat-headed tadpoles cling to rocks in flowing water or live in accumulated debris in backwaters. *Pseudobufo asper*, the sole member of its genus, is the only aquatic species in the Bufonidae and lives in

LEFT Natterjack toad, *Epidalea calamita* (formerly *Bufo calamita*), Ainsdale, Lancashire, UK.

RIGHT Long-fingered slender toad, *Ansonia longidigitata*, Sabah.

RIGHT Lowland dwarf toad, *Pelophryne signata*, Sarawak.

peat swamps in parts of Borneo, Sumatra and peninsular Malaysia. It has fully webbed hind feet and calls from floating mats of vegetation. The tree toads, *Pedostibes*, with four species from Southeast Asia and a fifth from India, are arboreal species that have longer limbs and digits than terrestrial bufonids but are otherwise similar in appearance. They have been found up to 6 m (20 ft) from the ground in canopy trees, but descend to the ground to breed in the backwaters of lowland rivers and rocky streams. They are probably the only arboreal members of the family apart from two small toads belonging to the genus *Nectophryne*, from West Africa. Staying in Africa, two members of the genus *Nectophrynoides*, *N. tornieri* and *N. viviparus*, are ovo-viviparous and the single species of *Nimbaphrynoides*, *N. occidentalis*, (sometimes considered to be two separate species, *N. liberiensis* and *N. occidentalis*) is viviparous.

South American stream frogs, Hylodidae

Three genera, *Crossodactylus*, with 11 species, *Hylodes* with 24 species and *Megaelosia* with 7 species, make up this family, which is restricted to Brazil and northern Argentina. The first two genera contain small diurnal frogs that live along rocky forest streams, with the *Crossodactylus* species being more aquatic in their habits than the other two, which enter the water less often. *Megaelosia* species grow up to 120 mm (4 $^7/_{10}$ in) whereas the *Hylodes* species are smaller. All species breed aquatically and their tadpoles live in the same streams as the adults. *Hylodes asper* has been shown to communicate with other frogs of the same species by waving its hind feet, a behaviour known as foot-flagging, and which is also recorded in unrelated species such as the *Staurois* species from Borneo, which lives in similar habitats, and *Micrixalus saxicola*. It seems quite possible that other species of *Crossodactylus* behave in a similar way.

Cryptic forest frogs and rocket frogs, Aromobatidae

The Aromobatidae consists of one hundred species in five genera formerly included within the Dendrobatidae, or poison dart frogs. Unlike the members of that family, however, they do not have the ability to produce skin toxins from their prey. As a result, they lack the bright warning colours seen in many poison dart frogs and are, instead, brown and cryptically coloured, often with a broad dark band along their flanks. Some species have bright undersides or throats.

Apart from one species, the Venezuelan skunk frog, Aromobatids live on forest floors where their coloration makes them difficult to see. Although not every species has been studied, they are diurnal and lively, constantly on the move. Many of them lay small clutches of eggs on the ground and one of the parents remains nearby. When they hatch, the parent transports them on its back to a small forest pool or quiet backwater in a river or stream, where they continue to develop aquatically.

There are three subfamilies. The Aromobatinae contains two genera, *Aromobates* and *Mannophryne*, with 12 and 16 species, respectively. They live in northern South America and almost every species is endangered or critically endangered. One species, the Venezuelan skunk frog, *Aromobates nocturnus*, gives off a distinctive smell, likened to the odour of a skunk, despite its lack of toxins. This species is unusual in other ways; it is green in colour with a row of yellow spots on its flanks and it is larger than other species in the family. It is also nocturnal and has only ever been found sitting or floating in water and so it is more aquatic than other members of the family. This is one of the critically endangered species.

The Anomoglossinae also contains two genera, *Anomoglossus* and *Rheobates*, with 22 and two species, respectively. They are all small brown frogs of the forest floor where their coloration makes them difficult to see. Most species appear to lay their eggs on dead leaves and the parents carry the tadpoles to water but the small yellowish-brown

Anomaloglossus beebei (*Colostethus beebei*) breed in large terrestrial bromeliad plants. The eggs are laid on a leaf above the waterline and the parents keep them moist until they hatch, at which point the male carries them to the water-filled centre of the plant. The female subsequently lays infertile eggs into the water to feed them.

The Allobatinae contains 47 species in a single genus, *Allobates*. Most species are cryptically coloured but the brilliant-thighed frog, *A. femoralis*, which sometimes has orange or tan limbs, are more colourful. At least two *Allobates* species produce non-feeding tadpoles. *Allobates degranvillei* carries its tadpoles until they metamorphose whereas the tadpoles of *A. stephani* develop in a terrestrial nest.

Poison dart frogs, Dendrobatidae

The 174 species in this family are divided into 12 genera and three subfamilies. Their taxonomy is very complicated and the relationships between them are not fully understood. Many names have changed in the last decade. A further 100 species were recently placed in a separate family, the Aromobatidae. Poison dart frogs live in the northern half of South America and adjacent parts of Central America. Many are brightly coloured and produce skin toxins of the most potent kind. South American Indians use the secretions to tip their blow-gun darts, giving them their common name (although only a handful of species are actually used in this way). They are active, diurnal and have fascinating life-histories. There could hardly be a more charismatic group of frogs.

They are mostly rainforest species, living on the forest floor or in shrubs and low-growing trees. Some species live around water courses but they are, more or less, independent of free-standing water. Their breeding habits show a high degree of

RIGHT An unusual colour form, blue and black, of the green and black poison dart frog, *Dendrobates auratus*.

parental care. The eggs are laid on the ground, usually on a dead leaf or other surface that the male has cleaned. Some species hide the eggs under an overlapping leaf, or lay them in the centre of a curled-up leaf. There is only a cursory amplexus although courtship varies between the species. Once the eggs have been laid and fertilized, one or other of the parents guards them until they hatch, at which point they wriggle onto the parent's back. They are then carried to a small body of water, often in the central tank of a bromeliad plant, but the adults also use the broken husks of forest fruits and nuts and other receptacles. An introduced population of *Dendrobates auratus* on Hawaii apparently uses discarded tins and broken bottles.

The subfamily Colostethinae contains mostly cryptically coloured species although some of the *Ameerga* species have brightly coloured stripes down their bodies, which are otherwise black, and some have dark red backs. Several have brighter colours on their throats and undersides and some of the six *Epipedobates* species have cream or lime green stripes on a reddish background. Some members of this subfamily have toxic alkaloids in their skin secretions but they are not considered to be among the most poisonous species in the family, and chemicals from *E. tricolor* have been investigated for possible use as pain-killers.

BELOW Strawberry poison dart frog, *Oophaga pumilio*, Costa Rica.

The Dendrobatinae contains many brightly coloured species. The five *Phyllobates* species are all brilliantly coloured and their skin toxins are among the most potent. The golden poison dart frog, *P. terribilis*, is thought to be the most poisonous animal in the world. Owing to recent name changes the genus *Dendrobates* now includes only five species but the blue poison dart frog, *D. tinctorius azureus*, is arguably the most colourful of all frogs. The nine *Oophaga* species have elaborate life cycles in which the female returns to the pools where she has deposited the tadpoles and provides them with an infertile egg to eat. (*Oophaga* means egg-eater.) The best-known species is *O. pumilio*, the strawberry poison dart frog, while another, the harlequin poison dart frog, *O. histrionica*, exists in a myriad of different colours and patterns that can make identification difficult. The largest genus, *Ranitomeya*, consisting of 27 species, also includes species with a similar breeding system to the *Oophaga* species, although parental care in most of them stops short of supplying their tadpoles with eggs. The three brightly coloured *Adelphobates* species deposit their tadpoles in the empty husks of Brazil nuts.

The third subfamily, Hyloxalinae, contains 57 species in the genus *Hyloxalus*. Many are cryptically coloured but some species, *Hyloxalus azureiventris*, for instance, have a pair of brightly coloured stripes passing down either side of their backs, which are otherwise black or brown. This particular species also has blue markings on its underside. Where known, the parents of these species carry their tadpoles to small natural pools or to the backwaters of larger streams and rivers.

Narrow-mouthed frogs, Microhylidae

This is a very large family with 466 species divided into 54 genera and many genera are small in number or have only a single species. They are often divided into subfamilies, of which there are about 11, depending on which system is followed, although several genera have not been positively assigned to a subfamily. It seems likely that some of the subfamilies will be elevated to full families in the future. Members of the Brevicipitidae are sometimes included in this family (as a subfamily) but here they are treated separately. Collectively, they have a wide distribution, in North, Central and South America, Africa, Madagascar, Asia, including New Guinea, the Philippines, Indonesia and Australia. Subfamilies tend to be more restricted in their range with, for instance, the Dyscophinae, Scaphiopryninae and the Cophylinae confined to Madagascar, and the Phrynomerinae to the southern half of Africa.

Many microhylids are small and secretive, and consequently difficult to study, but some are arboreal and a small proportion are large and colourful. Arboreal species have expanded toe discs. Calling usually takes the form of chirps, peeps and trills while some sound like the bleating of distant sheep.

Reproduction is varied throughout the family. Most Asian and American species produce large numbers of small eggs that float on the surface of puddles and temporary pools. In many cases, their tadpoles are mid-water filter-feeders. Others produce aquatic tadpoles that develop without feeding, one of them, *Syncope antenori*, from South America, using the water that collects in terrestrial bromeliads. Further steps towards independence of water is taken in many members of the family, such as the 15 *Kalophrynus* species from Southeast Asia and all 50 members of the Cophylinae from Madagascar, which lay their eggs in damp or rain-filled depressions in the ground or in rotting logs. The eggs hatch into tadpoles that remain in their nests and complete their development without feeding. Direct development occurs in a number of species, including, as far as is known, all of those from New Guinea, Australia and neighbouring regions (subfamily Asterophryinae) several of which are arboreal. Some South American species are also thought to have direct development, as do some Madagascan species, but the natural history of most microhylids is poorly known, or not known at all, so there could be exceptions.

The Asterophryinae is a large subfamily found from the Philippines, through Indonesia to New Guinea and northwestern Australia. All the Australian species are restricted to the rainforests of northeastern Queensland. Most species are small, rotund frogs with short limbs. They live under the ground or among leaf-litter and males call from a hidden location. They lay their eggs in depressions in moss or damp soil and, in some but perhaps all species, the males remain with them, presumably to guard against predation or parasitism by small invertebrates, until they develop into small froglets, skipping the aquatic stage altogether. The Australian *Cophixalus* species are sometimes known as nursery frogs because males guard the eggs after they are laid. The eggs are large in size but small in number, and unpigmented.

The Cophylinae is restricted to Madagascar, where 50 species are divided into seven genera. They may be arboreal and the members of several genera (*Anodonthyla*, *Cophyla* and *Platypelis*) breed in water-filled tree-holes in tropical forests. Terrestrial species, *Madecassophryne*, *Rhombophryne* and *Stumpffia*, lay their eggs among leaf-litter, either in small depressions or in foam nests. *Plethodontohyla* includes both tree-hole and terrestrial breeders and eggs of both types hatch into tadpoles that remain in the nest and do not feed. Males guard the nests in at least some of the species.

The Dyscophinae is also Madagascan and has three species all in the same genus, *Dyscophus*. They are popularly known as tomato frogs although only one species, *D. antongilii*, is bright red in colour. They are terrestrial and *D. antongilii* lives in gardens in towns and villages. They breed in backwaters, ditches and swamps.

The Gastrophryninae includes 45 species in nine genera, from North and South America. They are small robust species with pointed snouts, living in grasslands, scrub and rainforests, usually in burrows or hidden among forest debris. Some have bleating calls from which they get their common name of sheep frogs while members of the large genus *Chiasmocleis*, with 23 species from Central and northern South America, are called humming frogs. Two out of the five species of *Gastrophryne* occur in the United States and one other, *Hypopachus variolosus*, is a Central American species that just reaches southern Texas. As far as anyone knows, all the members of this subfamily lay large rafts of small eggs that float on the surface of pools or swamps.

ABOVE Female tomato frog, *Dyscophus antongilii*, Madagascar. Males of this species are duller in colour than females. They occur only in and around the town of Maroansetra, on the Bay of Antongil, where they breed in ditches and stagnant ponds.

The Hoplophryninae contains only three species in two genera, *Hoplophryne* and *Parhoplophryne*, from the Eastern Arc Mountains of Tanzania. They live in montane forests and are arboreal, breeding in tree-holes and bamboo stems.

The Kalophryninae consists of a single genus, *Kalophrynus*, with 15 species from Southeast Asia. They live on forest floors among leaf-litter and are known as sticky frogs due to the latex-like substance they secrete from their skin if they are handled. They have pointed snouts and pear-shaped bodies and there is often a well-defined line where their dark back meets their lighter-coloured flanks, making them hard to see when they are resting among dead leaves. They are ant- and termite-eaters and one of the more common species, the rufous-sided sticky frog, *K. pleurostigma*, feeds by stationing itself alongside ant-trails. Some species lay their eggs in small bodies of water that collect in fallen logs, while others breed in small water-filled depressions, pools and swamps. The tadpoles of at least some species do not feed.

The Melanobatrachinae contains but a single species, *Melanobatrachus indicus*, from the Western Ghats of southern India, where it lives in the vicinity of small streams. Nothing is known of its natural history.

The Microhylinae has 69 species in nine genera, all from Southeast Asia. They are mostly terrestrial species with typical robust bodies, living in forests, grasslands and urban habitats. Thirty species belong to the genus *Microhyla*, the largest and most widespread of the subfamily. They are small with narrow heads and very long hind legs. They are all terrestrial, and breed in small pools or rain puddles, laying a raft of small eggs that float on the surface, except the two species of *Metaphrynella*, which are tree-hole breeders. The Asian painted bullfrog, *Kaloula pulchra*, is one of the larger species, and is common in and around villages. It forms large breeding aggregations in drains and ditches after heavy rain and under buildings, the males calling with loud, plaintive, or moaning notes. The seven members of the genus *Calluella* are known as spadefoot frogs owing to the horny flange on their hind feet, which they use to burrow into the soil. In this they parallel the spadefoot toads of North America, Europe and Australia, none of which are closely related, of course. The saffron-bellied frog, *Chaperina fusca*, has bright yellow spots on its underside, and the colour is said to rub off onto human fingers.

The Otophryninae contains a single genus, *Otophryne*, with three species from northern South America. Tadpoles of *O. pyburni* live between grains of sand on stream bottoms and feed on tiny organisms that dwell there.

The Phrynomerinae contain five species in a single genus, *Phrynomantis*, known as rubber frogs, from the southern half of Africa. They have bright red or orange markings on a shiny black background and the red-banded rubber frog, *P. bifasciatus*, and probably the other species too, produce skin toxins. They are medium-sized frogs with elongated bodies that live in grasslands and semi-desert regions. *Phrynomantis annectens* lives in the Namib Desert and breeds in deep waterholes in granite outcrops but most species congregate to breed in flooded pans or shallow pools, including those that form in elephants' footprints. The tadpoles are filter feeders and those of *P. annectens* and *P. bifasciatus*, and possibly others, form large shoals or clumps but scatter if they are disturbed.

The Scaphiophryninae has two genera and ten species, and is endemic to Madagascar. Two species of *Paradoxophyla* are small and typical microhylids, with pear-shaped bodies and pointed snouts. They both live in rainforests and are terrestrial. The eight *Scaphiophryne* species are more rotund in shape and have rounded snouts. Some live in rainforests and may be arboreal, but others occur in dry habitats, and are burrowers, coming to the surface only to breed in temporary pools. The remarkable *S. gottlebei* is among the most colourful microhylids and lives in a limited area of sandstone canyons in the Isalo National Park, in south-central Madagascar, where it hides in crevices during dry weather and breeds explosively in temporary pools after rains. Breeding in the burrowing species has rarely been observed but is thought to occur explosively at the beginning of the rainy season. Several species are threatened by collection for the pet trade.

LEFT The rainbow frog, *Scaphiophryne gottlebei*, Isalo, Madagascar. This is a colourful microhylid from a small area of sandstone canyons in Isalo National Park, Madagascar.

African rain frogs, Brevicipitidae

This is a small family of 26 species in five genera. Rain frogs occur in southern and eastern Africa, from the Cape Region to Ethiopia. They are rotund in shape, with a short head, hence the family name deriving from 'brevi-ceps', and short limbs. They inflate their bodies if they are picked up, making them even more spherical. Rain frogs burrow backwards and the species living in dry regions spend the greater part of their lives underground, emerging on wet nights to feed and breed. In order to achieve amplexus, the male, which is significantly smaller than the female, secretes a sticky substance from his chest, effectively gluing himself to the female's back. When she burrows back down into the ground she takes him with her and the eggs are laid and fertilized in an underground chamber. They develop directly, skipping the free-living tadpoles stage altogether.

The largest and best-known genus is *Breviceps*, with 15 species from dry places in southern and East Africa. Most are brown, buff or tan in colour but the strawberry rain frog, *B. acutirostris*, is flushed with red or pink. The commonest and most wide-ranging species is *B. adspersus*, which occurs in southwest Africa and extends up into the northern Namib Desert and Angola. The habits of forest species from East Africa, belonging to the genera *Callulina*, *Probreviceps* and *Spelaeophryne*, are poorly known as they all have limited ranges in the remote mountains of Tanzania and Zimbabwe, while the Ethiopian short-headed frog *Balebreviceps hillmani*, was only discovered in 1986 and is known only from the type locality in the Bale Mountains of Ethiopia.

BELOW Namaqua rain frog, *Breviceps namaquensis*, Western Cape, South Africa.

Shovel-nosed frogs, Hemisotidae

There are nine shovel-nosed frogs, all in the genus *Hemisus*, and they occur in tropical and subtropical Africa south of the Sahara. They are small to medium-sized frogs, with a distinctive pointed snout and rotund, swollen body. They burrow head first into damp substrate, using their hard, pointed snout to drive a burrow through the ground and using their feet to push soil out of the way. Males and females pair up on the surface at the beginning of the rainy season, and they remain in amplexus while the female digs a burrow adjacent to a temporary pool, in which to spawn. After the eggs are laid, which may be up to 2,000 in some species, the female remains with them until they hatch, at which point she digs out of the side of the nest chamber, releasing the tadpoles into the pool. Once in open water they develop in the normal way and metamorphose in three to four weeks. There is some evidence that the tadpoles swarm over the female's back and are partially carried out of the nest chamber, while other observations seem to show that the tadpoles are released as a result of heavy rains washing away the soil separating the chamber from the main pool, or even that they continue to develop in the jelly mass. No doubt there is variation between the species but the natural history of most of this family is poorly known.

ABOVE **Marbled shovel-nosed frog**, *Hemisus marmoratus*.

Squeakers and bush frogs, Arthroleptidae

The Arthroleptidae is an African family occurring south of the Sahara. There are 139 species in eight genera and they are divided into two subfamilies: the Arthroleptinae, with 88 species in seven genera, and the Leptopelinae, with 51 species in the genus *Leptopelis*. Members of the former are quite variable and difficult to describe in general terms. Forty-three members of the genus *Arthroleptis*, or squeakers, live in isolated patches of forest and have high-pitched calls. They lay small clutches of large eggs among leaf-litter and they develop directly into froglets. Other genera lay aquatic eggs that go through a tadpole stage in the normal way. The genus *Trichobatrachus* contains only one species, the hairy frog, *T. robustus*. This species is terrestrial but breeds in fast-flowing rivers and streams. During the breeding season the male develops a row of hair-like skin projections, or papillae, on its flanks, which are thought to increase the surface area of the frog and thereby aid in respiration underwater.

Members of the genus *Leptopelis* are known as bush frogs because they often live in low vegetation. They have previously been placed in various other families, including the Hyperoliidae. They possess adhesive toe pads that vary in size according to the habits of the species concerned, some being more arboreal than others. The toad-like *L. bufonides*, lives in grassland and lacks toe pads altogether. They have stocky bodies,

ABOVE Uluguru bush frog, *Leptopelis uluguruensis*, Tanzania.

very large, forward-facing eyes, and tend to be brown or fawn, although some are green or marbled green and brown. The long-toed bush frog, *L. xenodactylus*, which occurs in the Drakensberg Mountains of South Africa and is terrestrial, lives in marshes and calls from tussocks of grass, or from burrows in the saturated ground. All species, where known, lay eggs in depressions in damp ground. The tadpoles are large, with long tails, and make their own way to water by wriggling across the ground.

Reed, sedge and leaf-folding frogs, Hyperoliidae

The Hyperoliidae consists of 208 species in 18 genera. One genus, *Heterixalus*, with 11 species, is endemic to Madagascar and another, *Tachycnemis*, with a single species, is endemic to the Seychelles Islands. The other species are all African, and are found south of the Sahara wherever there is suitable habitat. They are restricted to wetlands or areas with ample water such as ponds and reed beds, where they live on emergent vegetation and, in some species, on trees and shrubs at the water's edge. They all have expanded toe pads for climbing and clinging to vegetation. Some species rest in exposed positions, even on hot sunny days, and appear to be resistant to desiccation. All have aquatic tadpoles but some species spawn on leaves and stems overhanging water whereas others lay their eggs in water, attached to aquatic plants. The tadpoles have long tails and high fins, and are light brown in colour. Patterns and colours of the adults vary tremendously within the family and many species are polymorphic, with similar colour forms overlapping whole groups of species; the *Hyperolius viridiflavus* species complex is particularly difficult with very many regional colour forms gradually merging from one to another. The argus reed frog, *Hyperolius argus*, is sexually dimorphic, with pale green males and brown and cream spotted or striped females.

The largest genus is *Hyperolius* with 128 species whose range covers that of the entire family on the African mainland. Wherever they occur they are usually present in large numbers, and even ornamental pools in gardens, hotel grounds, etc. often have a resident population. The males call in the evening and into the night, with what is often a repetitive single note. Breeding activity continues as long as the weather is warm and humid, which can be almost throughout the year in some parts of the range. Leaf-folding frogs, *Afrixalus*, of which there are 31 species, are similar to the *Hyperolius* species but have vertical pupils and small spines on the skin of their backs, especially in males. Most

ABOVE Marbled reed frog, *Hyperolius marmoratus*, Tanzania.

LEFT Pygmy leaf-folding frog, *Afrixalus brachycnemis*, Tanzania.

ABOVE Female white-spotted
reed frog, *Heterixalus
alboguttatus*, Madagascar.
Males of this species are often
uniform in colour.

are marked with brown and cream stripes but some have blotches and spots. Their legs and toes are long and slender and they are agile climbers. Some are very small, growing to only 20 mm (⅘ in). As their name suggests, leaf-folding frogs lay their eggs on leaves above the water. The leaves are then folded over and glued together by a secretion from the oviduct. Each pair can make several such nests in a single night. The *Kassina* species are known as running frogs and they are not as arboreal as the genera already described. They are more heavily built and either lack toe pads altogether or they are reduced in size. Most are cream with black longitudinal stripes but the red-legged kassina, *K. maculata*, is blotched in shades of brown and has an area of bright red on its thighs. The single species of *Semnodactylus*, the rattling frog, *S. weali*, is similar to the striped *Kassina* species.

The Madagascan reed frogs, *Heterixalus*, are similar to the African reed frogs, being arboreal, nocturnal, brightly coloured and, in some cases, polymorphic. The common *H. madagascariensis*, for example may be white, yellow or blue in colour. They breed in small bodies of still water and the eggs and tadpoles are aquatic. The single reed frog from the Seychelles, *Tachycnemis seychellensis*, is larger than most reed frogs, with females growing to 76 mm (3 in). It is an arboreal species that breeds in ponds and has aquatic tadpoles.

Species complexity in reed frogs

The reed frogs of Africa belonging to the genus *Hyperolius* seem to be going through a state of speciation. They all look rather similar in size and shape, but differ in subtle factors, so some species are impossible to tell apart from exterior appearance and certain identification can only be made by analysing their calls, examining a large series of specimens, plotting their distribution or by DNA analysis. Some former 'species' have recently been shown to include one or more cryptic species in the light of recent studies.

With the exception of *H. parkeri*, and a small group of species allied to *H. nasutus*, all species occur in two phases. These have been described as phase J, which applies to all juveniles and many males, and phase F, which applies to mature females and some mature males. Some phase J frogs develop into phase F adults, other remain as phase J but can be either males or females. Only frogs showing phase F can be easily identified as all phase J individuals tend to look the same, regardless of species. Phase J frogs are usually brown or green with a pair of pale lines down either side of their backs.

Phase F frogs are often very colourful and bear little or no resemblance to their phase J counterparts. The spiny-throated reed frog, *H. spinigularis*, differs in having two morphs that do not conform to the F and J phases described above.

The complexities do not stop here. Some species have many different colour and pattern forms, even as phase F. Sometimes these differences correspond with geographical locations but superimposed on this is variation even within the same population. *Hyperolius viridiflavus* for example, occurs over a wide area of East and southern Africa and approximately ten colour forms, or 'morphs' have been named. Some of these are referred to as a separate species, *H. marmoratus*, by some authorities and this in turn includes several morphs. Another species-group includes many morphs of *H. parallelus* and *H. marginatus*, which may in reality be the same species.

All this serves to show that frog speciation is not static and that selective pressures are still shaping populations. Species do not always fit into the hard and fast schemes that we would sometimes like them to.

African grass frogs, Ptychadenidae

There are 53 species in this family, all but four of them placed in the genus *Ptychadena*, the others being *Hildebrandtia*, with three species and *Lanzarana*, with one. *Ptychadena* species are long-legged, sharp-snouted frogs with distinctive ridges of skin running down their backs. Several species have a single pale-coloured line down their backs and some are dimorphic, with a proportion of the population having the line and others being plain. They live in open grassland and marshes and move about in dense vegetation, making them difficult to see and even more difficult to catch. They are also capable of enormous leaps, easily clearing more than 1 m (40 in) and often making three or four jumps in rapid succession. The Mascarene grass frog, *P. mascareniensis*, has a wide range over much of Africa south of the Sahara and has also been introduced to the Mascarene Islands, Madagascar and the Seychelles. All species breed in the still water of swamps, marshes, drainage ditches, etc., laying eggs in several small batches. The ornate frogs, *Hildebrandtia*, are somewhat similar but more heavily built. They are marked in

ABOVE Mascarene grass frogs, *Ptychadena mascareniensis*, from Madagascar occur in striped and unstriped forms in roughly equal numbers.

green and, sometimes tan. They have a large crescent-shaped spade on their hind feet and spend much of their lives below ground in open, dry grasslands, emerging after heavy rains to breed in temporarily flooded pans. The remaining species, *Lanzarana largeni*, lives in the arid scrub of Somalia and parts of Ethiopia and is poorly known.

The triangle frog and its relations, Ceratobatrachidae

Eighty-four species are grouped into the Ceratobatrachidae although their relationships are not fully understood and future changes may be made. They occur in Southeast Asia, including the Philippines and the Solomon Islands. They are mostly terrestrial but include some arboreal species. As far as is known, they all lay terrestrial eggs that have direct development.

The 68 *Platymantis* species are widespread within the region covered by the family as a whole. They include arboreal species with large toe pads as well as stocky, terrestrial species. Eggs are laid on the ground or in the leaf axils of plants. Five *Discodeles* species are similar but some are semi-aquatic and eight *Batrachylodes* species are also similar but generally smaller. *Palmatorappia solomonis* is the sole member of its genus and comes from the Solomon Islands. It is arboreal and, in appearance, it is superficially similar to hylid tree frogs, being brightly coloured in green or yellow and with wide toe pads on all its digits. The triangle frog, also known as the Solomon's leaf frog, *Ceratobatrachus guentheri*, is the most distinctive species in the family, being roughly triangular in outline with fleshy horns over its eyes and snout. It is variable in colour and may be brown, reddish or mustard yellow, all colours that make it hard to see when resting among dead leaves on the forest floor. It bears a strong resemblance to the Borneo horned frog, *Megophrys nasuta*, to which it is not related. It has a wide gape and eats large items of prey, including other frogs and even smaller members of the same species.

RIGHT Solomon Islands leaf frog or triangle frog, *Ceratobatrachus guentheri*.

Indian torrent frogs, Micrixalidae

The Micrixalidae contains only one genus, *Micrixalus*, containing 11 species, formerly part of the Ranidae. They occur in southwestern India, and are mostly from the Western Ghats mountains, or surrounding hills. They live along torrents, often resting on wet rocks in, or bordering, the stream. At least one species, *M. saxicola*, uses its hind feet to signal to other males entering its territory. They attach their eggs to rocks but their natural history is otherwise poorly known.

Indian leaping frogs, Ranixalidae

Ten species in the genus *Indirana* are currently placed in a family of their own, Ranixalidae. They occur in central and southern India and were formerly included in the Petropedetidae and, prior to that, the Ranidae. They are ranid-like in appearance and similar to the *Micrixalus* species, living in forests near rivers and streams. Their tadpoles apparently have elongated, flattened bodies and suctorial discs on their undersides, allowing them to graze over rocks in fast-flowing streams. Otherwise, nothing is known of their natural history.

African puddle frogs, Phrynobatrachidae

All 80 members of this family are contained in a single genus, *Phrynobatrachus*, which occur throughout Africa south of the Sahara, except the driest parts of the Namib Desert. They were previously included in the Ranidae.

ABOVE An African puddle frog, *Phrynobatrachus mababiensis*, from northern Namibia.

They occur in a variety of habitats wherever there is standing water, including open grasslands, rainforests, cultivated fields and irrigation and drainage ditches. They are small, plump brown frogs, somewhat warty or with folds of raised skin on their backs. Some species have yellow or orange mid-dorsal stripes and others are variable in this respect. Puddle frogs are very successful, breeding continuously in almost any body of still water. Their eggs are small and float in a single layer at the surface and they develop quickly. They can reach sexual maturity within five months and so two generations each year are possible under ideal conditions.

African water frogs, Petropedetidae

This family includes two genera, *Conraua* and *Petropedetes*, with six and 12 species respectively. They occur in Africa south of the Sahara and live alongside flowing streams that pass through rainforests. They are highly aquatic, although the natural history of many species is poorly known. *Petropedetes* species are small to medium-sized stream-dwelling frogs whose tadpoles, where known, are elongated and live on the surface of rocks in streams. *Conraua* species are highly aquatic with flattened bodies and heads, large hind feet and protruding eyes. They have granular skin and are dull brown or greenish in colour. Most species are poorly known. *Conraua derooi*, has a very limited range in Togo and is listed by the IUCN as critically endangered. *Conraua goliath* is the Goliath frog, or giant river frog, the largest frog in the world, growing to 33 cm (13 in) in length and weighing 3 kg (6 ⅗ lb) or more. It is dull green in colour with a yellow underside. Its large hind feet are fully webbed and it is a powerful swimmer. This species lives along cascades and torrents in Cameroon and Equatorial Guinea in West Africa and lays several hundred eggs in a mass, attached to vegetation at the bottom of rivers in the vicinity of rapids. Its tadpoles have a restricted diet for the first few days and only eat a certain plant, *Dicraea warmingii*, which grows on the rocks in suitable habitats. The Goliath frog's range is fragmented due to habitat destruction from logging and agricultural development, and it is widely eaten by local people, who consider it a delicacy.

African bullfrogs, sand frogs, stream frogs and their relatives, Pyxicephalidae

The members of the Pyxicephalidae take many forms and it is difficult to generalize about them and the validity of the family is not universally accepted. They occur in Africa south of the Sahara and were previously included in the Ranidae. They are divided into two subfamilies.

The Cacosterninae contains 63 species in 11 genera. Fifteen species of *Amietia* are known as river frogs and were previously placed in the genus *Rana*. They are large frogs with pointed snouts and powerful hind limbs, brown or greenish in colour. Some individuals have

BELOW Clicking stream frog, *Strongylopus grayii*, South Africa.

a yellow vertebral stripe. They live on river banks and near other large bodies of water and the Drakensberg stream frog, *A. vertebralis* is highly aquatic. This and some other species have limited ranges in the Drakensberg Mountains of South Africa, where they live in streams that may freeze in winter. Members of the genus *Cacosternum* are known as cacos or dainty frogs. The 12 species occur in southern and East Africa and include two described recently. They are small frogs with elongated bodies, slender limbs and no webbing between their toes. Some species, such as the common caco, *C. boettgeri*, are highly polymorphic, with individuals of many colours and patterns being found together. They breed in flooded fields, swamps, ditches and streams and their tadpoles develop rapidly. *Tomopterna* are sand frogs or pyxies, found in grassland and scrub, including arid regions, over much of the southern half of Africa. They are stout-bodied, burrowing species with a spade on their hind feet, enabling them to burrow backwards into the soil. They often breed in temporary bodies of water after rains but may also use more permanent breeding sites, including artificially constructed water holes. The stream frogs, *Strongylopus*, are similar to the river frogs but smaller. They live alongside streams and rivers and breed in quiet bodies of water, although the clicking stream frog, *S. grayii*, lays its eggs out of water in damp soil and the tadpoles make their way to the water after they have hatched. The eggs of the Namaqua stream frog, *S. springbokensis*, are laid on damp ground near water and do not develop until rains cause the level to rise. The chirping frogs, *Anhydrophryne* and *Arthroleptella*, lay their eggs in moss or damp leaf-litter and they either hatch into non-feeding tadpoles or develop directly. The micro frog, *Microbatrachella capensis*, found only in a small area around Cape Town, is critically endangered due to housing developments. The remaining four genera, which each contain a single species, are *Ericabatrachus*, *Natalobatrachus*, *Nothophryne* and *Poyntonia*.

The subfamily Pyxicephalinae contains only five species, two in the genus *Aubria*, which come from the rainforests of West Africa, and three species of *Pyxicephalus*, *P. adspersus* and *P. edulis*, both of which are widespread throughout most of southern and eastern Africa, and a third, *Pyxicephalus obbianus*, which comes from Somalia. These are large, heavily-built frogs that can live in dry grasslands and breed in temporary pools and flooded pans. The African bullfrog, *P. adspersus*, has a wide gape and tackles large prey. Males of this species guard their tadpoles and fend off potential predators, and will also dig a channel to deeper water if the tadpoles are in danger of being cut off by falling water levels.

BELOW African dwarf bullfrog, *Pyxicephalus edulis*.

Fanged frogs, grass frogs, seep frogs and their relatives, Dicroglossidae

This family, previously included in the Ranidae, is divided into two subfamilies, the Dicroglossinae and the Occidozyginae. Collectively, they range from Africa, through the Middle East to South Asia, the Philippines, Borneo and New Guinea.

BELOW Smooth guardian frog, *Limnonectes palavanensis*, Sabah, Borneo.

Members of the Dicroglossinae are ranid-like in appearance, and may be terrestrial, semi-aquatic or highly aquatic. There are 148 species in 12 genera, the largest of which is *Limnonectes*, with 53 species from India, through Southeast Asia to Japan. These are ubiquitous semi-aquatic frogs from a range of habitats, including streams, seeps, swamps and cultivated areas. Some species lay their eggs in scrapes that they make at the side of streams or in mud and the males stay with the eggs until they hatch. In the two guardian frogs, *L. finchi* and *L. palavanensis*, newly hatched tadpoles climb onto the male's back and he transports them to water. The males of some species from Southeast Asia, sometimes known as fanged frogs, and including *L. kuhlii* and *L. malesianus*, have fang-like structures on their lower jaw, presumed to be for territorial combat. Large species such as these eat other frogs. Thirty-two species of *Fejervarya* occur from Pakistan to Southeast Asia. They are sometimes known as grass frogs and are generalists, living in almost any damp situation, including flooded fields and rice paddies and males of *F. limnocharis*, for example, even call from puddles and potholes in tarred roads. The crab-eating frog, *Fejervarya cancrivora*, tolerates brackish

RIGHT Short-headed burrowing frog, *Sphaerotheca breviceps*, Puttulam, Sri Lanka.

ABOVE Spotted puddle frog, *Occidozyga lima*, Southeast Asia.

water and hunts for food on tidal mudflats and this large species, in turn, is eaten by humans. The 28 *Nanorana* species are rather similar to *Fejervarya* but are found further north, into the Himalayas and southern China. The five *Sphaerotheca* species are known as bullfrogs or burrowing frogs and occur in India and Sri Lanka. Their shape is similar to the spadefoot toads of Europe and North America and they possess a spade-like tubercle on their hind feet, for burrowing. Tiger frogs in the genus *Hoplobatrachus*, have a disjunct distribution, with *H. occipitalus* occurring in northeast Africa and the other three species in Asia. These are large frogs that will eat smaller frogs and, like the crab-eating frog, are eaten by humans. Two species have been introduced beyond their natural ranges in the vain hope that they would provide a source of food. The rugose frog, *H. rugulosus*, with a natural range throughout southern China and adjacent parts of Southeast Asia, has been introduced to Borneo, while the tiger frog, *H. tigerinus*, from India and neighbouring countries, has been introduced to Madagascar. The tadpoles of all four *Hoplobatrachus* species are carnivorous.

The other subfamily, Occidozyginae, has two genera, each with 11 species. The *Ingerana* species are small and stocky, and live mainly in forests. Their natural history is poorly known. The seep, or puddle frogs, *Occidozyga*, are small, stocky and have short limbs. They are commonly found in small bodies of shallow, muddy water, where their pale coloration makes them hard to see. Their tadpoles are elongated and live on the bottom of small pools and feed on aquatic invertebrates.

Mantellas and their relatives, Mantellidae

The Mantellidae contains 186 species in 12 genera. They are divided into three subfamilies, the Boophinae, containing only the genus *Boophis* with 70 species, the Laliostominae, containing four species in two genera and the Mantellinae, with the remaining nine genera and 112 species. Until recently, only two genera, *Mantella* and *Mantidactylus* were recognized within the Mantellinae, but recent studies have helped to explain the relationships between members of the former *Mantidactylus* species and led to the creation of eight separate genera. This does not tell the whole story, however, as there are several species awaiting description and, quite probably, more awaiting discovery. Mantellids are only found in Madagascar and the nearby Comoros Islands, where they form the dominant family of frogs.

Boophis species, sometimes known as bright-eyed frogs, have enlarged toe pads and most are arboreal, although a few, notably *Boophis microtympanum* and related species, occur at high altitudes where the habitat is open savannah, and the toe pads are not as well developed in these species. Arboreal species often have large eyes and may be green or brown in colour. Brown species, such as *Boophis madagascariensis*, often have small projecting flaps of skin on their ankles. Green species lack these appendages but have

BELOW Red-eyed boophis, *Boophis luteus*, Andasibe, Madagascar.

LEFT Madagascar bullfrog, *Laliostoma labrosum*, Ampijoroa, Madagascar.

large, brightly coloured eyes: red in *B. luteus*, yellow in *B. rappiodes* and blue-rimmed in *B. viridis*, for example. Males tend to be much smaller than females in most species and call from low vegetation. They lay their eggs in streams or stagnant forest pools and the tadpoles are aquatic.

The three members of the Laliostominae are terrestrial. *Laliostoma labrosum* is sometimes called the Madagascan bullfrog. It lives in the drier forests in the west of the country and spends the dry season underground, emerging to breed after heavy rain. The other genus in the subfamily, *Aglyptodactylus*, consists of three named species, and at least three more yet to be formally described. They are all quite similar and used to be known collectively as *A. madagascariensis*. They occur on the floor of rainforests, and are brownish in coloration, making them difficult to see. Males of some species, however, develop yellow patches of skin during the breeding season. All lay their eggs in temporary ponds.

The remaining species, placed within the Mantellinae, are more varied. They range in size from tiny species such as *Wakea madinika*, males of which grow to only 13 mm (½ in), and the slightly larger *Blommersia kely*, to *Mantidactylus guttulatus*, which grows to 100 mm (4 in) or more. They are found in a range of habitats, including fast-flowing streams, forest litter and trees. Arboreal forms have expanded toe pads whereas terrestrial species do not. Most are relatively slender and have long legs but a few are more robust-looking. Certain *Spinomantis* species have fringes of pointed, fleshy spines on their limbs and bodies. The coloration of the frogs in this subfamily tends to follow their chosen habitat, with forest-floor species being mostly brown and arboreal ones green, but members of the genus *Mantella* are famous exceptions, being brilliantly coloured, diurnal, toxic frogs closely paralleling the American poison dart frogs, Dendrobatidae.

RIGHT Golden mantella, *Mantella aurantiaca*, with freshly laid eggs.

Reproduction among the mantelline frogs is unusual though not uniform across all genera. There is no amplexus. In arboreal species, males position themselves above females, while clinging to vertical leaves overhanging water. In all cases, where known, the eggs are laid out of water and the tadpoles make their way to water in various ways. In most *Mantella* species the eggs are laid in depressions in the ground which later flood after rain. The green-backed mantella, *M. laevigata*, however, lays its eggs in holes in trees or in bamboo sections and females return to the cavity to produce infertile food eggs for the larvae once they have hatched and have slid down into the water. Another group of species, belonging to the genus *Guibemantis*, live in large pandanus plants and lay their eggs in the leaf axils. The elongated tadpoles are able to flick themselves over the leaf surfaces to move from one axil to another. Finally, the large genus *Gephyromantis* contains species that undergo direct development from eggs laid in leaf-litter, as well as species that lay eggs in a variety of terrestrial situations but which eventually hatch into tadpoles that make their way to water and develop with or without feeding.

Asian tree frogs, Rhacophoridae

The Rhacophoridae has 319 species divided into two unequal subfamilies, the Buergeriinae with four poorly known species from Taiwan and Japan, and the Rhacophorinae, with 315 species from Asia and Africa. They are sometimes known as Asian tree frogs although not all of them are Asian, and nor are they all tree frogs. Some authorities include them in the Ranidae.

The nominate genus, *Rhacophorus*, contains 80 species that are often large and colourful, and which have heavily webbed feet. Several of these are flying frogs that live much of their lives in the forest canopy but glide down to lower levels to congregate around pools to breed. Well-known species include Wallace's flying frog, *R. nigropalmatus*,

Reinwardt's flying frog, *R. reinwardtii* and the harlequin flying frog, *R. pardalis*. All of these are from the rainforests of Southeast Asia but the genus as a whole extends from India to China and Japan. Many are predominantly green in colour in keeping with their arboreal habits. The 23 *Polypedates* species are similar and used to be classified as *Rhacophorus* but they lack webbing between their toes and are therefore unable to glide. *Polypedates* species are commonly brown in colour. Members of both these genera typically lay their eggs in a nest of foam attached to vegetation overhanging a small pool or, occasionally, on the mud bank of such a pool. Other breeding sites can include artificial bodies of water such as drainage ditches, water tanks and swimming pools.

LEFT Harlequin flying frog, *Rhacophorus pardalis*, Danum Valley, Borneo.

LEFT Pied warty tree frog, *Theloderma asperum*, Vietnam.

The largest genera in the family are *Philautus* and *Pseudophilautus*, with 48 and 106 species respectively, variously known as bush frogs or bubble-nest frogs. These genera of small, usually brownish species occur from India and Sri Lanka (where many new species have been named in recent years), across Southeast Asia and on to the larger island groups of the South Pacific. Bush frogs live among low vegetation and in leaf-litter and lay their eggs in moss or other damp substrates. Development is direct in all the species about which we have information.

Theloderma species are covered in warts and small pointed tubercles. They are known collectively as bug-eyed frogs or rough-skinned tree frogs, and all 17 species are Asian, with several discovered recently in the rainforests of northern Vietnam. They are masters of camouflage and most are predominantly brown in colour and rest on bark, but the mossy frog, *T. corticale*, is greenish. Breeding habits are not well known for all species but *T. corticale* and *T. asperum* both lay their eggs out of water, attached to the inside surfaces of tree-holes, so that the tadpoles drop or wriggle down into the water when they hatch. These two species are highly vocal and have a variety of calls, made by both males and females, for purposes that have not yet been established.

Three *Nyctixalus* species are reddish brown in colour with pearly white spots all over their bodies. Adult cinnamon frogs, *N. pictus*, spend most of their time in the forest canopy but apparently lay their eggs in water-filled holes in dead trees and logs.

The grey tree frogs of the *Chiromantis* genus are represented by three species in Africa and another 13 in Southeast Asia. The African grey tree frog, *C. xerampelina*, is common in arid grasslands of East Africa and breeds in waterholes and other small bodies of water. Its foam nests are attached to overhanging branches and their outer surface quickly dries to a hard crust, protecting the eggs from drying out. This species is one of only a handful of frogs that excretes uric acid, a white paste that requires almost no water to carry it out of the body, a water-saving system more commonly associated with birds and reptiles. The remaining 22 species are divided between five additional genera and most were previously included in either *Philautus* or *Rhacophorus*. As relationships between some of the lesser-known species are investigated, more species will probably be moved between genera.

The other subfamily, Buergeriinae, contains four species in the genus *Buergeria*, from Taiwan and the Ryukyu Islands, Japan. They are small terrestrial species that occupy a range of habitats but rarely climb.

Robust frogs, Nyctibatrachidae

This small family from India and Sri Lanka contains 17 species, 16 in the genus *Nyctibatrachus*, all from the Western Ghats of southern India, where they live along hill streams, and a single species of *Lankanectes*, the Sri Lankan wart frog, *L. corrugatus*. This species is unusual as its back is covered by a series of transverse folds, like an old-fashioned washboard. It spends its time in shallow water and the *Nyctibatrachus* species are also aquatic.

The natural history of many of the species, some of which are endangered, is unknown. *Nyctibatrachus petraeus* has a very unusual method of egg-laying. Despite being largely aquatic, spawning takes place on leaves overhanging streams. Males call to attract females and when a female arrives the male moves to one side. The female lays a small cluster of eggs on the leaf where the male was calling from, but there is no amplexus. When she has finished laying, the male moves back over the eggs and fertilizes them. He continues to call and may attract other females to his leaf. In the meantime, the first female moves off and may lay more eggs on another male's leaf. The lack of amplexus, coupled to males and females dividing their reproductive effort between several different partners, is very unusual, perhaps unique, among frogs and it would be interesting to see if the other *Nyctibatrachus* species have similar breeding systems.

LEFT Sri Lankan wart frog, *Lankanectes corrugatus*, Gampola, Sri Lanka.

True frogs, Ranidae

Formerly a large family containing several groups of species that did not have a common ancestor, the 'old' Ranidae has been divided into seven new families, the Dicroglossidae, Micrixalidae, Nyctibatrachidae, Phrynobatrachidae, Ptychadenidae, Pyxicephalidae and a much reduced 'new' Ranidae.

The Ranidae, as now understood, contains 342 species in 16 genera and has an almost worldwide distribution, absent only from the southern half of South America, most of Australia and the whole of New Zealand. Ranids are medium to large frogs, the American bullfrog, *Lithobates catesbeiana*, being the largest species at 180 mm (7 in). Most are associated with water although temperate species may wander far away from it outside the breeding season. The genus *Rana*, with 49 species, occurs throughout Europe, right across Asia and into Indo-China, and it is also represented in western North America. *Lithobates* species, of which there are 48, occur in North, Central and South America and include many species formerly placed in *Rana*. Members of both these genera breed in ponds, ditches, marshes and the borders of lakes, laying large clumps containing hundreds of eggs. Tadpoles of temperate species, such as the European common frog and the American bullfrog, may take more than one year to develop, over-wintering below the ice if their ponds become frozen over.

BELOW Common frog, *Rana temporaria*, Mull, Scotland.

LEFT Spotted stream frog, *Hylarana picturata*, Danum Valley, Sabah, Borneo.

A number of tropical species, such as those belonging to the Asian genera *Amolops* and *Meristogenys*, both of which are known as torrent frogs, breed in fast-flowing montane streams and rivers. Their tadpoles have large suckers on their undersides with which they cling to rocks, and they often work their way above the water level to graze on the emergent surfaces of water-lapped rocks. The 51 members of the large genus *Odorrana* are also cascade-dwellers and have strong-smelling skin that may be toxic, for example *Odorrana hosii* is known as the poisonous rock frog in Borneo and the Malaysian Peninsula. *Hylarana* species form the largest genus, with 86 species, and occur in northeast Africa and East Africa, and in south Asia from India and Sri Lanka to southern China, into Southeast Asia, the Philippines, New Guinea and northern Australia. They live in a variety of habitats, including rainforests and urban areas, and breed in temporary ponds or quiet backwaters of forest streams, where their tadpoles hide among accumulated dead leaves. The Australian wood frog, *H. daemeli*, is the only ranid in Australia. Five *Staurois* species live near waterfalls and torrents and are sometimes known as splash frogs or rock frogs. Two species from Borneo, the rock skipper, *S. lateropalmatus*, and the spotted rock frog, *S. natator*, extend their brightly coloured hind feet to signal to other frogs in a kind of semaphore, possibly because their calls cannot be heard above the sound of rushing water.

Twenty-two species of *Pelophylax* occur through most of Europe and into North Africa, the Middle East, Central and southern Asia. They include an interesting group of frogs that sometimes hybridize, giving rise to offspring that can go on to form self-perpetuating forms, and which are known as kleptons, from the Greek word for thief (because the hybrid form steals some of its chromosomes from one of its parent species). Strictly speaking, their scientific names should indicate this by the abbreviation kl., as in *Pelophylax* kl. *esculenta* (the edible frog), which is a form derived from the pool frog *P. lessonae* and the marsh frog, *P. ridibundus*. (All these frogs were previously placed in the genus *Rana*.) The remaining eight genera are from China and Southeast Asia and contain relatively few species. The natural history of many of them is poorly known.

Further information

References to recent classification (pp.16 and 17)

Frost, D. R., Grant, T., Faivovich, J., Bain, R. H., Haas, A., Haddad, C. F. B., De Sa, R. O., Channing, A., Wilkinson, M., Donnellan, S. C., Raxworthy, C. J., Campbell, J. A., Blotto, B. L., Moler, P., Drewes, R. C., Nussbaum, R. A., Lynch, J. D., Green, D. M. and Wheeler, W. C. 2006. The amphibian tree of life. *Bulletin of the American Museum of Natural History* 297: 1–370.

Frost, Darrel R. 2010. Amphibian Species of the World: an Online Reference. Version 5.4 (8 April, 2010). Electronic database accessible at http://research. amnh.org/vz/herpetology/amphibia/American Museum of Natural History, New York, USA.

Vitt, L. J. and Caldwell, J. P. 2009. *Herpetology*. Academic Press, Burlington, San Diego and London.

The most relevant titles are listed below.
A more comprehensive bibliography can be found in several of the large reviews.

Arnold, N. and Ovenden, D. 2002. *A Field Guide to the Reptiles and Amphibians of Britain and Europe.* HarperCollins, London.

Carruthers, C. 2001. *Frogs and Frogging in South Africa.* Struik, Cape Town. (Includes a CD with 56 frog calls.)

Cei, J. M. 1980. Amphibians of Argentina. *Monitore Zool. Ital.* (Italian Journal of Zoology) Mongr. 2, 1980.

Channing, A. 2001. *Amphibians of Central and Southern Africa.* Cornell Univeristy Press, New York.

Collins, J. P. and Crump, M. L. 2009. *Extinction in Our Times.* Oxford University Press, NY.

Conant, R. and Collins, J. T. 1998. *A Field Guide to Reptiles & Amphibians of Eastern and Central North America.* Peterson Field Guide Series, Houghton Mifflin Co., Boston.

Corkran, C. C. and Thoms, C. 2006. *Amphibians of Oregon, Washington and British Columbia.* Lone Pine Publishing, Edmonton, Vancouver, and Auburn, WA.

Duellman, W. E. 2001. *The Hylid Frogs of Middle America* (expanded edn.). Society for the Study of Amphibians and Reptiles (SSAR), Ithaca, NY.

Duellman, W. E. and Trueb, L. 1994. *Biology of Amphibians.* The Johns Hopkins University Press, Baltimore and London.

Elliott, L, Gerhardt, C and Davidson, C. 2009. *The Frogs and Toads of North America.* Houghton Mifflin Harcourt, Boston and New York. Includes a CD of frog calls.

Glaw, F. and Vences, M. 2007. *A Field Guide to the Reptiles and Amphibians of Madagascar* (third edn.). Vences and Glaw Verlag, Cologne.

Halliday, T. and Adler, K. (eds) 2002. *The New Encyclopedia of Reptiles and Amphibians.* Oxford University Press, Oxford.

Inger, R. F. and Stuebing, R. B. 2005. *A Field Guide to the Frogs of Borneo* (second edn.). Natural History Publications (Borneo), Kota Kinabalu.

Kwet, A. 2009. *European Reptile and Amphibian Guide.* New Holland, London, Cape Town, Sydney and Aukland.

Maeda, N. and Matsui, M. 1990. *Frogs and Toads of Japan* (second edn.). Bun-Ichi Sogo Shuppan, Tokyo.

Menzies, J. 2006. *The Frogs of New Guinea and the Solomon Islands.* Pensoft, Sofia and Moscow.

Myers, C. W., Daly, J. W. and Malkin, B. 1978. A dangerously toxic new frog (*Phyllobates*) used by Emera Indians of Western Colombia, with discussion of blowgun fabrication and dart poisoning. *Bulletin of the American Museum of Natural History* 161(2): 307–366.

Myers, C. W., Daly, J. W. and Martinez, V. 1984. An Arboreal Poison Frog (*Dendrobates*) from Western Panama. *American Museum Novitates,* 2783: 1 – 20.

Passmore, N. I. and Carruthers, V. C. 1995. *South African Frogs: a complete guide* (revised edn.). Witwatersrand University Press, Johannesburg.

Rödel, M–O. *Herpetofauna of West Africa. Vol 1. Amphibians of the West African Savanna.* Edition Chimaira, Frankfurt am Main.

Schiotz, A. 1999. *Treefrogs of Africa.* Edition Chimaira, Frankfurt am Main.

Stebbins, R. C. 1985. *A Field Guide to Western Reptiles and Amphibians* (of the United States). Peterson Field Guides, Houghton Mifflin Co., Boston and New York.

Stebbins, R. C. and Cohen, N. W. 1995. *A Natural History of Amphibians.* Princeton University Press, Princeton NJ.

Tyler, M. J. 1989 *Australian Frogs: a Natural History.* Cornell University Press, Ithaca and London.

Tyler, M. J. and Knight, F. 2009. *Field Guide to the Frogs of Australia.* CSIRO Publishing, Collingwood, Vic.

Vitt, L and Caldwell, J. P. 2009. *Herpetology* (third edn.). Academic Press, Burlington MA, San Diego, CA and London.

Wells, K. D. 2007. *The Ecology and Behavior of Amphibians.* University of Chicago Press, Chicago and London.

WEBSITES

Amphibian Specialist Group, publishers of "FrogLog" an online magazine relating to amphibians and amphibian conservation. http://www.amphibians.org/ASG/Home.html

AmphibiaWeb: Information on amphibian biology and conservation. [web application]. 2010. Berkeley, California: AmphibiaWeb. Available: http://amphibiaweb.org/.

ARKive: a centralised digital library of films and photographs of threatened species. http://www.arkive.org/Frogs Australia Network. http://www.frogsaustralia.net.au/

Frost, Darrel R. 2010. Amphibian Species of the World: an Online Reference. Version 5.4 (8 April, 2010). Electronic Database accessible at http://research. amnh.org/vz/herpetology/amphibia/ American Museum of Natural History, New York, USA.

Haas, A. & Das, I. (2010) Frogs of Borneo. The frogs of East Malaysia and their larval forms. Web resource accessible at: http://www.frogsofborneo.org/.International Union for the Conservation of Nature. http://www.iucn.org/

Partners in Amphibian and Reptile Conservation (PARC). http://www.parcplace.org/Pond expert. http://www.pondexpert.co.uk/

Index

Page numbers in *italic* refer to illustration captions; those in **bold** refer to main subjects of boxed text.

PICTURE CREDITS

© NHMPL pp. 6, 10. © Daniel Heuclin/NHPA/Photoshot p.19. © Minden Pictures/FLPA p.21 Tom and Pam Gardner; p.39 Pete Oxford; p.47 Dietmar Nill; p.58 bottom Chris & Tilde Stuart; p.68 Mark Moffett; p.76 Scott Leslie; p.77 top Paul Sawer; p.100 bottom Piotr Naskrecki; p.120 top © Erica Olsen; p.153 Thomas Marent; p.154 Wil Meinderts; pp.86 bottom, 96 top, 102 bottom, 134, 136, 147, 157, Michael & Patricia Fogden © Dr. Wolf Fahrenbach/Visuals Unlimited/Corbis p.22. © Naturepl.com p.28, 78 Doug Wechsler; p.93 top Mark Payne-Gill; p.148 Michael D. Kern. © Jean Paul Ferrero/Ardea. com p.61 bottom, p.62 right. © Photolibrary. com p.73 Wayne Lynch; p.81 Juniors Bildarchiv; p.120 bottom Stefan Huwiler; p.131 David M Dennis. © Bill Leonard p.84 bottom. © Vance Vredenburg p.128. © Atherton de Villiers p.142. © Karthickbala p.143. © Miguel Vences and Frank Glaw p.175. All other images © Chris Mattison.